Stabilization of Superconducting Magnetic Systems

THE INTERNATIONAL CRYOGENICS MONOGRAPH SERIES

General Editors

Dr. K. Mendelssohn, F. R. S.
The Clarendon Laboratory
Oxford, England

Dr. K. D. Timmerhaus
Engineering Research Center
University of Colorado, Boulder, Colorado

H. J. Goldsmid
  Thermoelectric Refrigeration, 1964

G. T. Meaden
  Electrical Resistance of Metals, 1965

E. S. R. Gopal
  Specific Heats at Low Temperatures, 1966

M. G. Zabetakis
  Safety with Cryogenic Fluids, 1967

D. H. Parkinson and B. E. Mulhall
  The Generation of High Magnetic Fields, 1967

W. E. Keller
  Helium-3 and Helium-4, 1969

A. J. Croft
  Cryogenic Laboratory Equipment, 1970

A. U. Smith
  Current Trends in Cryobiology, 1970

C. A. Bailey
  Advanced Cryogenics, 1971

D. A. Wigley
  Mechanical Properties of Materials
  at Low Temperatures, 1971

C. M. Hurd
  The Hall Effect in Metals and Alloys, 1972

E. M. Savitskii, V. V. Baron, Yu. V. Efimov,
M. I. Bychkova, and L. F. Myzenkova
  Superconducting Materials, 1973

W. Frost
  Heat Transfer at Low Temperatures, 1975

I. Dietrich
  Superconducting Electron-Optic Devices, 1976

V. A. Al'tov, V. B. Zenkevich, M. G. Kremlev, and
V. V. Sychev
  Stabilization of Superconducting Magnetic Systems,
  1977

Stabilization of Superconducting Magnetic Systems

V. A. Al'tov, V. B. Zenkevich,
M. G. Kremlev, and V. V. Sychev

Edited by
V. V. Sychev

Translated from Russian by
G. D. Archard

Translation Editor
K. D. Timmerhaus

With a Foreword by
B. W. Birmingham

PLENUM PRESS • NEW YORK AND LONDON

61542982

Library of Congress Cataloging in Publication Data

Main entry under title:

Stabilization of superconducting magnetic systems.

(The International cryogenics monograph series)
Translation of Stabilizatsiia sverkhprovodiashchikh magnitnykh
sistem.
Includes bibliographical references and index.
1. Superconducting magnets. I. Al'tov, Valeriĭ Aleksandrovich. II.
Series.
QC761.3.S713   621.39   77-8618
ISBN 0-306-30943-2

Physics

The original Russian text, published by Energiya in Moscow in 1975, has been
corrected by the authors for the present edition. This translation is published
under an agreement with the Copyright Agency of the USSR (VAAP).

СТАБИЛИЗАЦИЯ СВЕРХПРОВОДЯЩИХ МАГНИТНЫХ СИСТЕМ
В. А. Альтов, В. Б. Зенкевич, М. Г. Кремлев, и В.В.Сычев.

STABILIZATSIYA SVERKHPROVODYASHCHIKH MAGNITNYKH SISTEM
V. A. Al'tov, V. B. Zenkevich, M. G. Kremlev, and V. V. Sychev

© 1977 Plenum Press, New York
A Division of Plenum Publishing Corporation
227 West 17th Street, New York, N.Y. 10011

All rights reserved

No part of this book may be reproduced, stored in a retrieval system, or transmitted,
in any form or by any means, electronic, mechanical, photocopying, microfilming,
recording, or otherwise, without written permission from the Publisher

Printed in the United States of America

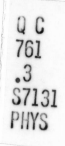

Q C
761
.3
S7131
PHYS

Foreword

I am indeed pleased to prepare this brief foreword for this book, written by several of my friends and colleagues in the Soviet Union. The book was first published in the Russian language in Moscow in 1975.

The phenomenon of superconductivity was discovered in 1911 and promised to be important to the production of electromagnets since superconductors would not dissipate Joule heat. Unfortunately the first materials which were discovered to be superconducting reverted to the normal resistive state in magnetic fields of a few hundredths of a tesla. Thus the development that was hoped for by the early pioneers was destined to be delayed for over half a century.

In 1961 the intermetallic compound Nb_3Sn was found to be superconducting in a field of about 200 teslas. This breakthrough marked a turning point, and 50 years after the discovery of superconductivity an intensive period of technological development began.

There are many applications of superconductivity that are now being pursued, but perhaps one of the most important is super-conducting magnetic systems. There was a general feeling in the early 1960s that the intermetallic compounds and alloys that were found to retain superconductivity in the presence of high magnetic fields would make the commercialization of superconducting magnets a relatively simple matter. However, the next few years were ones of disillusionment; large magnets were found to be unstable, causing them to revert to the normal state at much lower magnetic fields than predicted. Projects failed as this technological difficulty became apparent.

v

3996

Discoveries and developments soon followed which resulted in
a composite superconductor. A composite superconductor is one
which combines the superconducting material with a large amount
of metal in the normal conducting state. The object is to provide
an electrical shunt to the superconductor which will handle the
current should the superconductor suddenly revert to the normal
state.

The composite superconductor brought with it a whole new
series of problems related to trade-offs among stability, efficiency,
and ac loss. A large part of the solution of these problems was
found in the development of conductors in which the superconducting
path consists of finely divided, transposed filaments.

The authors have done a real service to the scientists and
engineers who work with superconducting magnetic systems. They
have presented a systematic exposition of the various aspects of
the problem of stabilizing superconducting magnets. They give a
qualitative discussion of various stability problems and then
proceed to put the analysis on a solid quantitative basis.

I want to commend the authors for an excellent book which
addresses the problems arising in stabilizing composite superconductors. This book should become an essential reference work
for those designing and working with superconducting magnetic
systems.

B. W. Birmingham

National Bureau of Standards
Boulder, Colorado, U.S.A.
July 1977

Preface

One of the most important lines of advance in modern technology is the creation of various kinds of superconducting devices.
Superconductivity is exploited in a large number of fields: the
power and electrotechnical industries, high-energy physics, and
cosmic technology.

Superconductivity is at present most widely employed in the
creation of strong magnetic fields, using superconducting magnetic systems. The parameters of such systems are rapidly advancing;
devices with very large values of stored energy are now being
made. In large magnetic systems the stored energy may be measured in tens and hundreds of megajoules. When the superconductivity of the winding degenerates, unless special precautions are
taken, this energy is liberated as heat, and accidents may occur.
It is thus a vital problem to find means of reliably preventing such
emergencies in the operation of magnetic systems.

Complete reliability of a magnetic system may in principle be
achieved by using a combined superconducting wire containing a
large amount of metal in the normal state; this metal must have
a high electrical conductivity and must shunt the superconductor.
To put it quite simply, for currents not too greatly exceeding the
nominal value the supermagnetic system may then operate for a
limited time as an ordinary cryogenic system, the current passing
through the low-temperature normal metal. The mean current
density in such a winding is not very high, since the proportion of
superconductor within the cross section of the wire is quite low.
In order to construct such a system it is essential to possess specific information as to the conditions of heat transfer in the wind-

ing and the manner in which the current is displaced from the
superconducting to the normal portion of the conductor. However,
the problem is seriously complicated by the fact that space and
weight limitations often necessitate the use of windings with a
high mean current density. The introduction of a high proportion
of normal metal into the combined conductor is then severely re-
stricted, and a compromise has to be made in relation to the reliability of the final solution.

In constructing an efficient superconducting magnetic system
the most important questions include those of the actual degree of
stabilization required (i.e., the conditions under which an uncontrolled transition of the winding into the normal state is to be prevented), and the overall reliability of the system (i.e., the development of criteria in coil construction and the creation of protective devices for preventing damage to the system during its transition into the normal state, whatever the cause of the transition).
This book is devoted to a consideration of these vital practical
problems.

Information regarding various aspects of the problem of stabilizing superconducting coils is at present scattered over the
pages of a large number of journals, and this naturally creates
great difficulties for the specialist who requires a reasonably complete acquaintance with the state of the problem. This book constitutes the first attempt at a systematic exposition of the problems
which arise in stabilizing combined conductors. Particular attention is devoted to the method of thermal (cryogenic) stabilization.
Methods of internal stabilization also receive final detailed consideration.

The material of the book is largely based on the results of
the authors' own investigations carried out over the last few years
in the Applied Superconductivity Department of the Institute of
High Temperatures of the Academy of Sciences of the USSR, but
the pertinent literature has also been reviewed.

The book is chiefly intended for scientists and engineers spe-
cializing in the development, manufacture, and application of various kinds of superconducting magnetic systems, as well as students
in these fields; it may also be useful for specialists working in
related fields of science and technology (heat transfer in the cryo-

genic region, metallurgy of combined superconductors, and so on), and others interested in superconducting magnetic systems.

The critical comments of readers will be gratefully received.

V. Sychev

Contents

Part II

METHOD OF THERMAL STABILIZATION

Part III

COMBINED CONDUCTORS WITH INTERNAL STABILIZATION

Part I

SUPERCONDUCTIVITY
AND ITS APPLICATIONS

Chapter 1

The Future of Superconductivity
in Modern Technology

1.1. Introduction

Practical applications of superconductivity were attempted immediately after the discovery of this phenomenon by H. Kamerlingh-Onnes in 1911 [1]. Kamerlingh-Onnes proposed using a superconducting solenoid as an electromagnet, producing strong magnetic fields by virtue of the high currents attained; the electrical power requirements would be very low, being chiefly determined by the power required to maintain the low temperatures [2]. However, this apparently simple idea proved impossible to achieve in practice, and attempts at making such an electromagnet in 1913 ended in failure [3]. It was found that ordinary superconductors returned to the normal state in magnetic fields with an induction of (at most) a few hundredths of a tesla. This made it impossible to use superconductivity in practical technology. This situation remained the same for almost fifty years, until, at the beginning of the 60's, a number of alloys and compounds remaining in the superconducting state in strong magnetic fields (order of tenths of a tesla), with high currents passing through the superconductor, were discovered. The discovery of these materials (nonideal superconductors of the second kind) marked a revolution in the practical use of superconductors. From this time a vigorous development of scientific-research and experimental-design work has been pursued relative to the introduction of superconducting devices into technology. A large number of laboratories in the more highly developed countries have concentrated their efforts in investigations of this kind.

There are now very real prospects of using superconducting devices in a variety of fields of modern technology (power and electrical technology, experimental physics, cosmic and radio technology, computing procedures, and so forth).

The most promising directions in the technical applications of superconductivity may be summarized as follows:

1. Various kinds of superconducting magnetic systems.
2. Superconducting electrical transmission lines.
3. Superconducting logical and memory units for computers.
4. Superconducting resonators and parts for hf circuits.
5. Superconducting gyroscopes and accelerometers, superconducting frictionless bearings.

1.2. Superconducting Magnetic Systems

The greatest range of possible applications and the greatest degree of present development lies in the field of superconducting magnetic systems. This is quite easy to understand, since by comparison with ordinary electromagnets* superconducting magnetic systems have far better economic and operating characteristics, and in many cases provide the one and only acceptable solution to particular problems. Superconducting magnetic systems have the following advantages:

1. These systems require no ferromagnetic cores and tolerate far higher current densities in the winding than ordinary magnetic systems; they are one or two orders of magnitude lighter and far more compact.

2. The practically complete absence of Joule losses in the winding ensures an expenditure of electrical power on their intrinsic requirements one to two orders of magnitude lower than that of an ordinary system; the electrical power is in this case consumed almost solely in maintaining the helium temperature in the coil (i.e., in supplying the helium liquefier or refrigerator); ordi-

---

*An ordinary magnetic system is the name which we shall give to an electromagnet with a ferromagnetic (usually iron) core, and copper or aluminum windings.

nary magnetic systems also require large flows of cooling water in order to eliminate Joule heating.

3. Superconducting magnetic systems give considerably stronger magnetic fields by comparison with ordinary systems; whereas ordinary technological magnetic systems with ferromagnetic cores give magnetic fields with inductions of no greater than 2 T, for large superconducting systems a magnetic induction of the order of 4-6 T is now quite normal. It should nevertheless be noted that at the present time several quite unique laboratory electromagnets of the Bitter type exist (strong-current water-cooled electromagnets without any ferromagnetic core). These electromagnets create extremely strong fields in small volumes (up to 25 T). However, the use of these magnets involves a very high energy consumption, and they are accordingly only used for physical research.

Superconducting magnetic systems have additional advantages over ordinary systems in relation to certain special problems. For example, the compactness of superconducting magnetic systems enables us to produce a magnetic field with a much higher induction gradient than that of ordinary systems. This is extremely important, for example, when studying elementary particles with short lifetimes [6].

Apart from ordinary and superconducting magnetic systems, cryogenic (or cryoresistive) magnetic systems may also be made; the windings in this application are made from especially pure metals (copper, aluminum,* sodium) and work at temperatures of about 20 K. At these temperatures the specific electrical resistance of the metals in question is two or three orders of magnitude smaller than at room temperature, and the Joule losses in the windings diminish accordingly. This gives cryogenic magnetic systems considerable advantages over ordinary electromagnets from the point of view of their intrinsic consumption of electrical power. However, by comparison with superconducting magnetic systems the intrinsic power requirements of cryogenic systems are at least an order of magnitude higher [7].

Naturally, in solving the question as to the desirability of using magnetic systems of any particular type we must not only allow for the advantages in relation to intrinsic power consumption but also analyze the capital expenditure and the amortization component of the annual expenses.

A calculation of the total annual expenditure involved in the use of magnetic systems of various types (ordinary, cryogenic, and superconducting) shows that, for cur-

---

*Aluminum is a superconductor, but its critical temperature is extremely low (T_c = 1.2 K), so that at hydrogen temperatures it behaves as an ordinary metal.

rent densities of over 10^4 A/cm² in the windings, superconducting magnetic systems are better than any others [8].

One of the most important lines of attack at the present time is the creation of superconducting magnetic systems for installations converting heat directly into electrical power with the aid of magnetohydrodynamic generators.

It has been established that the MHD power installations of thermal power stations are only economically efficient if superconducting magnetic systems are used to create the magnetic field in the MHD generator. In addition to considerably smaller size and negligible intrinsic power consumption, superconducting magnetic systems have one further extremely important advantage over ordinary magnetic systems: far higher magnetic fields. Since the length of the MHD generator channel is inversely proportional to the square of the magnetic induction in the generator, an increase in induction will clearly make the generators smaller. Increasing the magnetic field to 4 T shortens the MHD generator channel by a factor of 4 (by comparison with the length of an ordinary magnetic system). It should be noted that on shortening the channel the thermal losses through the walls are reduced as well.

In recent years various countries have pursued intensive scientific-research and experimental-design work in connection with the creation of superconducting magnetic systems for MHD generators.

Superconducting magnetic systems of various types are finding wider and wider applications in experimental nuclear physics. Large superconducting solenoids are used as magnets for bubble chambers, elementary-particle accelerators, and plasma traps in thermonuclear investigations.

The use of superconducting deflectors and magnetic focusing systems in elementary-particle accelerators is extremely promising. In the ordinary mode of assembly such systems require large quantities of power for their intrinsic needs. The high inductions and sharp field gradients obtainable with superconducting systems greatly reduce the distances over which beam focusing may be effected.

Superconducting windings for powerful motors and generators are certainly promising aspects of future electrical engineering. Present electrical-generator construction is characterized by

a general tendency toward increasing the power obtained in a
single plant (at present the typical level is 1000 MW); it is difficult,
however, to envisage any further large increase in the unit power
of ordinary generators (comprising steel frames and copper windings). The problem of creating generators (and electric motors)
of greater powers may be solved by using superconducting coils.
The power of an electrical machine is, in fact, proportional to the
magnetic induction in the working region and the current density
in the armature. These parameters may be greatly increased by
using superconductors. For an induction of the order of 6-7 T
and a current density in the armature of the order of 10^4 A/cm^2,
a generator with superconducting coils will have a power two
orders of magnitude greater than an ordinary generator of the
same dimensions. At the present time work in this direction is
being vigorously pursued in a number of countries.

The creation of large superconducting solenoids working as
energy storage devices (rings), intended to cover times of peak
loading in power systems or to supply power in an emergency,
has recently been seriously considered. It is shown, for example,
in [9] and [10] that superconducting solenoids are economically
more efficient for these purposes than any other type of energy-
storage device. Design work is continuing in this direction. A
design for a 10^{13} J superconducting power-storage device is described in [10]. Small model superconducting storage rings have
already been made and successfully tested in a number of countries. Work is also continuing on the creation of superconducting
storage rings for use as sources of powerful short energy pulses.

Thus superconducting magnetic systems of various kinds are
now being widely introduced into technology (for detailed reviews
see [11, 12]). Ensuring the reliable, accident-free operation of
such systems has accordingly become an extremely important
problem.

Chapter 2

The Nature of Superconductivity

2.1. General Principles

⌊It is well known that superconductivity is a special state of certain pure metals and alloys in which the electrical resistance of the metal (or alloy) is equal to zero; substances exhibiting the superconducting state are called superconductors.

The transition of a superconductor from the normal state, characterized by a specific value of the electrical resistivity, into the superconducting state takes place on cooling below a certain critical temperature⌉$T_{c0} = T_c(B)$ at $B = 0$, where B is the magnetic field induction. Different superconductors have differing values of T_c; the critical temperatures of superconductors so far discovered lie between 0.012 K (tungsten) and approximately 23 K (solid solution of $Nb_3Al - Nb_3Ge$). ⌈At the critical temperature, the resistance of the superconductor falls sharply from a certain finite value to zero. The zero value of the resistance is the most important characteristic of the superconducting state.⌉

The second fundamental characteristic of the superconducting state (the Meissner effect) lies in the fact that an external magnetic field will not penetrate through a superconductor in the superconducting state. We may thus consider that for a superconductor in the superconducting state the magnetic permeability μ is equal to zero; in other words, such a conductor forms an ideal diamagnetic material.

The external magnetic field falls to zero in a thin layer at the surface of the superconductor. The equations of macroscopic electrodynamics (the London equations) lead to the following equa-

9

tion for the magnetic induction in a superconductor occupying a half-space x > 0 [13]:

$$B(x) = B(0) e^{-x/\lambda} , \qquad (2-1)$$

where x is the distance along the normal from the surface of the superconductor, and λ is the London depth of penetration.

For ordinary electron densities $\lambda \approx 10^{-6}$ cm; for certain superconductors (such as Nb$-$Zr alloy and the intermetallic compounds Nb_3Sn and V_3Ga) λ extends to 2×10^{-5} cm. The surface layer of the superconductor carries undamped superconduction currents; their average field compensates the external magnetic field so that the latter cannot penetrate into the interior of the superconductor. It follows from the first Maxwell equation that current can only flow in the surface layer of the superconductor in which $\partial B/\partial x \neq 0$.

If a superconductor in the superconducting state (i.e., at $T < T_c$) is subjected to a sufficiently strong external magnetic field, the superconductivity is destroyed, i.e., the magnetic field penetrates inside the superconductor and the latter returns to the normal state (despite the fact that $T < T_c$). The lower the temperature of the superconductor, the greater is the magnetic field required to destroy the superconductivity. This magnetic field is called critical and its induction* is denoted by B_c. The dependence of B_c on T for the majority of conductors is approximately parabolic and may be closely described by the empirical equation

$$B_c(T) = B_0 \left[1 - \left(\frac{T}{T_c} \right)^2 \right], \qquad (2-2)$$

where B_0 is the critical magnetic field induction at T = 0 K (determined by extrapolation).

Clearly the macroscopic state of a superconductor is uniquely determined by specifying its temperature and external magnetic field. Figure 2-1 shows the B$-$T diagram of a superconductor. The shaded region corresponds to the superconducting state, and the region above the curve to the normal state. The curve relating the critical magnetic induction of the field to temperature is the

---

*The constant B_0 in Eq. (2-2) equals 4.13×10^{-2} T for mercury, 3.045×10^{-2} T for tin, 8.05×10^{-2} T for lead, 2.69×10^{-2} T for indium, and 1.0×10^{-2} T for aluminum.

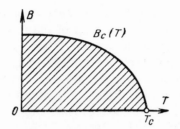

Fig. 2-1. The B—T diagram of a superconductor.

geometrical locus of points corresponding to the transition of the
superconductor from the superconducting into the normal state.

Superconductivity is destroyed when the critical magnetic
field is reached on the sample surface. Here it is immaterial
whether this field is created by an external magnet, by the current
flowing through the superconductor, or a combination of these two
(Silsby rule). Usually $B_c(T)$ is relatively small,* and no substantial current can pass through a superconductor of this type.
The magnetic field created by this current exceeds the critical
value, and the superconductor passes into the normal state.

For the majority of superconductors the transition from the
superconducting to the normal state is reversible — it does not
involve any irreversible loss of energy. This is a direct conse-
quence of the Meissner effect; it enables us to consider the superconducting and normal states as two phases and to apply the ordinary methods of thermodynamics in order to analyze the laws
governing the transition [14]. Superconductors in which the transition from the superconducting to the normal state in the presence
of a magnetic field is thermodynamically reversible are called
ideal superconductors.

The nature of superconductivity was established in 1957 by
Bardeen, Cooper, and Schrieffer, who developed the microscopic

---

*The definition of the critical magnetic field in terms of the induction of the field B_c
rather than the intensity H_c is accepted here in the interests of uniformity. Since in the
majority of cases the superconductor may be regarded as surrounded by a nonmag-
netic medium, this change is of a nominal character. It should be remembered that
in certain cases, for example, in that of a superconductor surrounded by magnetic
material, the transition will take place, not when the critical induction has been
reached in the latter, but when the critical field H_c has been attained.

theory of superconductivity [15]. In 1956 Cooper [16] found that under certain conditions (as a result of the interaction of the free conduction electrons with the crystal-lattice phonons of a conductor) bound states of the electrons might develop — these became known as Cooper pairs.

As a result of this interaction, a fluctuation arises in the positive ionic charge of the lattice, resulting in the dynamic screening of the electric field of the electron. The fluctuation in the positive charge may be so great that it overcompensates the Coulomb field of the electrons (i.e., "suppresses" their Coulomb repulsion) and leads to attraction between the electrons, i.e., to the formation of electron pairs. The Bardeen—Cooper—Schrieffer theory is based on the concept of these Cooper pairs.

The Cooper pairs have a whole-number spin; in contrast to the ordinary electron gas, which obeys Fermi—Dirac statistics, the Cooper pairs (like any particles with a whole-number spin) obey Bose—Einstein statistics. Whereas a Fermi gas cannot be condensed, a nonideal Bose gas may pass into a special state with falling temperature. This transition may be interpreted as a special kind of condensation in momentum space. When Cooper pairs are created, the conduction electrons "condense" into a state characterized by a greater degree of order.

It is well known that a condensed nonideal Bose gas (such as He II) possesses the property of superfluidity.

According to the Bardeen—Cooper—Schrieffer theory superconductivity may be regarded as superfluidity of the Cooper pairs.

Important results for analyzing the phenomenon of superfluidity were obtained by means of the phenomenological theory developed by Ginzburg and Landau [17].

It follows from the thermodynamic relationships describing the phase transition from the superconducting into the normal state that the entropy of the superconductor is lower in the superconducting state than in the normal. Thus the superconducting phase has greater internal order than the normal phase. A thermodynamic description of superconductivity may be presented even without analyzing the microscopic nature of the superconducting state by purely macroscopic considerations, making use of a certain order parameter ψ. It is an approach of this nature (analogous to the method used by Landau in his construction of the theory of phase transitions of the second kind) which is employed in the phenomenological Ginzburg–Landau theory.

The order parameter ψ is introduced in such a way that, if the superconducting and normal phases are in thermodynamic equilibrium, it changes from zero in the normal region of the superconductor to unity in that part of the superconducting region furthest removed from the normal region. If we consider, in conformity with the conclusions of the Bardeen–Cooper–Schrieffer theory, that the superconducting state is characterized by a mixture of superconducting (i.e., linked into Cooper pairs) and normal (i.e., single) electrons, the parameter ψ may be defined as the proportion of superconducting electrons in this mixture. In the Ginzburg–Landau theory (set up in 1950 before the appearance of the Bardeen–Cooper–Schrieffer theory, when the microscopic nature of superconductivity had still not been elucidated) the quantity ψ was defined as a certain parameter relative to which the free energy of the superconductor was expressed as a series expansion.

In the equations of the Ginzburg–Landau theory the most important part is played by the Ginzburg–Landau parameter

$$\varkappa = \lambda(T)/\xi(T),$$

(2-3)

where $\lambda(T)$ is the London penetration depth, satisfactorily described by the empirical equation

$$\lambda(T) = \frac{\lambda(0)}{\sqrt{1 - \left(\dfrac{T}{T_c}\right)^4}},$$

(2-4)

and $\xi(T)$ is the coherence length defined by the relation

$$\xi(T) = \xi_0 \left(\frac{T_c}{T_c - T}\right)^{\frac{1}{2}}.$$

(2-5)

The quantity ξ_0 represents the minimum distance at which the ordering parameter may vary, or in other words the smallest extent of the superconducting region surrounded by a normal region. The coherence length may also be considered as the size of a Cooper pair. For the majority of superconductors $\xi_0 \approx 10^{-5}\text{-}10^{-4}$ cm; for transition metals such as tantalum and niobium $\xi_0 \approx 10^{-6}$ cm; for Nb−Zr alloys and the intermetallic compounds Nb_3Sn and V_3Ga $\xi_0 \approx 5 \times 10^{-7}$ cm.

Abrikosov [18] achieved an approximate solution for the non-linear equations of the Ginzburg−Landau theory.* This solution enabled the fundamental laws governing the nature of the super-conducting state in superconductors of various types to be established.

2.2. Superconductors of the First and Second Kinds

[As indicated by Ginzburg and Landau [17], the structure of a superconductor in the superconducting state is determined by the sign of the surface energy at the interface between the super-conducting and normal phases. Thermodynamic consideration of the change in the free energy of the system at the interface between the phases (using the concepts of the coherence length ξ and depth of penetration λ) shows that the specific free energy of the interface is given by the equation

$$f_{\text{sur}} = (1 - \varkappa) \xi \frac{B_c^2}{2\mu_0},\qquad (2\text{-}6)$$

where $\mu_0 = 4\pi \times 10^{-7}$ ΓH/m is the magnetic constant.

It follows from this equation that, if $\varkappa < 1$ (i.e., $\xi > \lambda$), then $f_{\text{sur}} > 0$, and hence an increase in the surface of separation between the superconducting and normal phases is thermodynamically unfavorable for superconductors of this type] An increase in the area of the interface would increase the free energy of the system, and this is quite impossible in a spontaneous process [14]. The free energy of the system reaches a maximum when the area of the interface assumes the least possible value. It follows that for

---

* Gor'kov [19] derived the equations of the phenomenological Ginzburg−Landau theory from the equations of the microscopic Bardeen−Cooper−Schrieffer theory. The modern phenomenological theory of superconductivity linked to the microscopic theory is accordingly called the Ginzburg−Landau−Abrikosov−Gor'kov theory, after the names of the authors of Refs. [17-19].

superconductors in which $f_{sur} > 0$ the state with a minimum area of the interface between the phases is stable, and thus the penetration of normal zones into the superconducting region is impossible.

The situation is quite different for superconductors in which $\varkappa > 1$ (i.e., $\xi < \lambda$) and hence $f_{sur} < 0$. Clearly for superconductors of this type an increase in the surface of separation between the superconducting and normal phases is thermodynamically advantageous, since in this case the free energy of the system diminishes. It follows that for superconductors in which $f_{sur} < 0$ the state with an extensive phase interface is stable, i.e., one in which normal zones "impregnate" the superconducting region.

Ideal superconductors with a positive surface energy are usually called superconductors of the first kind; ideal superconductors with a negative surface energy are called superconductors of the second kind.

It should be noted that Abrikosov's more rigorous treatment leads to an expression for the surface energy differing slightly from (2-6); the surface energy in this case is positive for $\varkappa < 1/\sqrt{2}$ and negative for $\varkappa > 1/\sqrt{2}$.

The foregoing ideal superconductors, for which the Meissner effect occurs (the magnetic field falling to zero in a surface layer with a thickness of the order of λ) and in which superconductivity breaks down when subjected to a specific induction B_c, are superconductors of the first kind. These include "soft" metals such as mercury, tin, lead, and indium.

Among ideal superconductors of the second kind there are a number of alloys (for example, single crystals formed from Pb−Te and Pb−In alloys of particular compositions).

Let us consider the basic properties of ideal superconductors of the second kind. The penetration of the normal zones into the superconducting region (thermodynamically favorable for such superconductors) represents the penetration of the external magnetic field into the interior of the superconductor.

In 1950 F. London suggested that, if the temperature of a doubly connected sample of superconducting material were reduced to $T < T_c$ and the external magnetic field were then reduced to zero, the magnetic flux which would in this way be "captured" by the sample would only be able to assume discrete values [20]. This proposition was verified experimentally in 1961 by B. S. Diver and W. Fairbank, and independently by R. Dolley and M.

Nabauer [21]. It was thus established that the magnetic flux Φ in matter could only change discretely, so that

$$\Phi = m\varphi_0, \qquad (2-7)$$

here m is a whole number and φ_0 is the minimum value of the change in the magnetic flux (fluxoid or flux quantum), in which

$$\varphi_0 = \frac{hc}{2e} \approx 2 \times 10^{-15}, \quad \text{W}, \qquad (2-8)$$

where h is Planck's constant, c is the velocity of light, and e is the charge on the electron.

According to Abrikosov's results the external magnetic field penetrates into the interior of a superconductor of the second kind by way of the penetration of individual fluxoids. The superconductor is thus permeated by microscopic regions of cylindrical shape; within each of these the material remains in a normal state. These regions are called vortex filaments or Abrikosov vortices. The vortex filaments (vortices) are naturally directed parallel to the external magnetic field; every such vortex carries a magnetic flux equal to one fluxoid. The radius of a vortex equals the coherence length ξ. The magnetic flux φ_0 concentrated along such a filament is sustained by the vortical motion of the superconducting electrons around the filament. The structure of a vortex is illustrated schematically in Fig. 2-2.

Over a distance equal to the depth of penetration λ the inductance falls from the value of B at the center of the filament (equal to the induction of the external field) to B/e, where e is the base

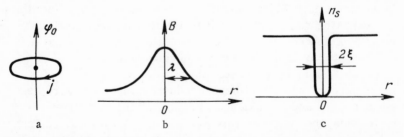

Fig. 2-2. Structure of a vortex or vortical filament. a) Vortical current j around a filament sustaining a fluxoid φ_0; b) and c) distribution of the magnetic field and the density of the superconducting electrons.

of natural logarithms. The radius of the vortical current is approximately equal to λ.

The penetration of the external magnetic field into the superconductor takes the form of filaments of normal phase penetrating the superconducting matrix. Abrikosov showed that the free energy of such a system was lowest when the vortices formed a regular triangular lattice.

The penetration of a vortex into the superconducting phase is not thermodynamically favorable for any arbitrary external magnetic fields, but only occurs on condition that*

$$B \geqslant \mu_0 \varepsilon / \varphi_0, \qquad (2-9)$$

where ε is the energy of unit length of the vortex, being given by the equation

$$\varepsilon = \frac{1}{4\pi\mu_0} \left(\frac{\varphi_0}{\lambda} \right)^2 \ln \frac{\lambda}{\xi}. \qquad (2-10)$$

Only for these values of B is the condition

$$dG_{\text{tot}} + dL^* \leqslant 0 \qquad (2-11)$$

satisfied, where dG_{tot} is the elementary change in the thermodynamic potential of the system on passing from the original state (in which the superconductor is entirely superconducting) to the state under consideration (in which the superconducting matrix is penetrated by the vortices); dL^* is the elementary useful work of the system (in the present case this is the energy of formation of a vortex).

It is well known from thermodynamics [14] that Eq. (2-11) is the condition for the equilibrium of a thermodynamic system when T = const, p = const, and H = const.

The magnetic field having an induction corresponding to the equality sign in (2-9) is called the first critical field

---

*Equation (2-9) is not entirely rigorous, since the interaction between the vortices under consideration and the neighboring filaments was not taken into account when deriving the equation; qualitatively, however, this inequality is valid.

$$B_{c_1} = \frac{\mu_0 \varepsilon}{\varphi_0} = \frac{\varphi_0}{4\pi\lambda^2} \ln \frac{\lambda}{\xi};$$ (2-12)

B_{c1} is a characteristic of superconductors of the second kind.

Usually for ideal superconductors of the second kind the first critical field is comparatively weak — a few thousandths or hundredths of a tesla [22].

It is clear from the foregoing that in fields weaker than B_{c1} the vortices will not penetrate into a superconductor at a temperature $T < T_{c0}$. Here the superconductor of the second kind behaves in the same way as superconductors of the first kind, in that the Meissner effect appears.

For $B > B_{c1}$ a finite density of these vortices occurs in a superconducting matrix existing in the equilibrium state for $T < T_c$. The superconductor is then in a mixed state: The external magnetic field penetrates into the regions existing in the normal state; yet electrical resistance is completely absent, since the rest of the volume of the superconductor is occupied by the superconducting phase.

The greater the external magnetic field, the higher is the density of the vortices; i.e., the greater proportion of the cross section of the superconductor is occupied by the normal phase. Finally, for a certain value of B at which the distance between the centers of the vortices becomes comparable with 2ξ practically the whole volume of the superconductor passes into the normal state.* The external magnetic field then penetrates into the superconductor and the magnetic moment vanishes. The magnetic field induction for which the magnetic moment of the superconductor vanishes is called the second critical induction (B_{c2}). It is obvious from elementary considerations (Fig. 2-2) that

$$B_{c2} \approx \varphi_0/4\xi^2.$$ (2-13)

Thus for $B > B_{c2}$ the superconductor is in the normal state. It should nevertheless be noted that, strictly speaking, when B_{c2} is exceeded the superconductivity is completely destroyed in the

---

*Superconductors with high values of $\varkappa$ (of the order of 10-100) pass into the normal state at fields lower than those associated with the closing of the vortices, because of the paramagnetic effect [23].

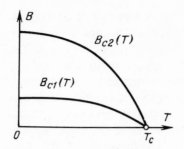

Fig. 2-3. The B–T diagram of a superconductor of the second kind.

main volume of the superconductor, but a thin layer (approximately 10^{-5} cm thick) remains superconducting on the surface of the
sample until the field obeys the inequality $B < B_{c3}$ where $B_{c3} \approx$
$1.69 B_{c2}$ [24].

The B–T diagram of a superconductor of the second kind is
schematically illustrated in Fig. 2-3; the region under the curve
$B_{c1}(T)$ corresponds to the completely superconducting state, the
region above the curve $B_{c2}(T)$ to the completely normal state, and
the region between these curves to the mixed state. The phase
transitions in an ideal superconductor of the second kind at $B = B_{c1}$
(transition from the superconducting to the mixed state) and at
$B \approx B_{c2}$ (from the mixed to the normal state) from the thermo-
dynamical point of view are phase transitions of the second kind.

It was noted earlier that, despite the absence of electrical
resistance, no large currents could be passed through a superconductor of the first kind. Let us now consider the question of
the possible passage of currents through an ideal superconductor
of the second kind.

To distinguish from the annular surface currents in superconductors of the first kind and the vortical currents around the vortices, we shall subsequently call the current in a superconductor
excited by an external source the transport current.

Whereas in a superconductor of the first kind the current is
only able to pass through the surface layer (i.e., the region in
which $\partial B/\partial x \neq 0$), in an ideal superconductor of the second kind,
subject to the condition $B_{c1} < B < B_{c2}$, for which the magnetic
field penetrates into the interior of the superconductor, the elec-

tric transport current is capable of flowing through practically the whole cross section not occupied by the vortices.

Clearly this transport current flowing through the superconductor will interact with the magnetic field, that is, with the vortices. Since the dimensions of the superconductor are finite, and hence the possibilities that a considerable spatial redistribution of the transport current may occur are limited, this interaction will lead to a displacement of the vortices under the action of a force F_L. The value and direction of this force are determined by the Ampere law, which for the case under consideration may be written in the form

$$F_L = \int_V \mathbf{J} \times \mathbf{B} \, dV, \qquad (2\text{-}14)$$

where $\mathbf{J}$ is the transport current density and V is the "volume" of a vortex.

Neglecting changes in current density at a distance of the order of the size of the vortex, we obtain

$$F_L = \varphi_0[\mathbf{J} \times \mathbf{n}]l, \qquad (2\text{-}15)$$

where $\mathbf{n}$ is the unit vector along the direction of the vortex and l is the length of this vortex (filament).

In papers relating to superconductors of the second kind the force F_L acting on the vortex is usually called the Lorentz force, although this term is not really quite justified, since the Lorentz force usually means the force acting on an electron moving at a velocity $\mathbf{v}$ in a magnetic field of induction $\mathbf{B}$, i.e., $F_{Lor} = e\mathbf{v} \times \mathbf{B}$.

Clearly, in the presence of a transport current, and subject to the condition $B > B_{c1}$, the magnetic field inside an ideal superconductor of the second kind, in which the "carriers" are vortices, is a moving field. This moving magnetic field leads to the development of an emf directed along the current (exactly as in the case of a moving current-carrying conductor in a stationary magnetic field). As a result of this, Joule heat is liberated, the temperature rises above the critical value for the field in question, and the superconductor passes into the normal state at fields much smaller than B_{c2}. Thus ideal superconductors of the second kind, like ideal superconductors of the first kind, are not suitable for transporting any substantial currents. [An exception is the case in which the direction of the transport current coincides with the direction of the magnetic field; in this case $F_L = 0$, as may readily be seen from (2-14) and (2-15).] Hence superconductors of these types cannot be used for creating superconducting structures carrying heavy currents.

A special class of superconductors is formed by nonideal superconductors of the second kind. These are similar to ideal superconductors of the second kind in that, for $B > B_{c1}$, they are in a mixed state, while for $B = B_{c2}$ (even in the absence of a current) they pass into the normal state. A distinctive characteristic of nonideal superconductors of the second kind is the fact that the crystal structure contains defects and inhomogeneities, so that in superconductors of this type a mechanism preventing the free motion of the vortices is provided for. The number of crystal-lattice defects, inhomogeneities in the composition, and so on, depends on the method of production, the composition, and the form of machining and heat treatment applied. These structural defects play the part of centers restraining the vortices; they may conveniently be called detaining (or restraining) centers.

The class of nonideal superconductors of the second kind includes various alloys and also a number of pure "hard" superconductors. The most important (from the point of view of practical use) nonideal superconductors of the second kind are listed in Table 2-1, together with the corresponding values* of T_c and B_{c2}.

TABLE 2-1

Superconductor		T_c, K	B_{c2}, T	
Composition	Structure		$T=0$ K	$T=4.2$ K
Nb_3Sn	Intermetallic compound (structure of the β—W type)	18.3	$\sim$30	22.1
V_3Ga	Ditto	16.8	$\sim$30	
Nb—25%Zr	Alloy	10.2		$\sim$7
Nb—60%Ti	"	9.0—9.5		$\sim$12
Nb—25%Zr—10%Ti	"	9.8—10.0		$\sim$10—11

---

*The field inductions B_{c1} of nonideal superconductors of the second kind (like their ideal counterparts) are relatively low; for example, in the case of V_3Ga at $T = 0$, $B_{c1} \approx 2 \times 10^{-2}$ T. The values of B_{c2} are given for a critical current density J_{c1} equal to zero.

We see from this table that B_{c2} for the superconductors indicated reaches extremely high values. It should be emphasized that the values of B_{c1} and B_{c2} for nonideal superconductors of the second kind depend only slightly on the number of defects and inhomogeneities in the crystal structure.

The most important characteristic of nonideal superconductors of the second kind, offering excellent prospects for their use in heavy-current devices, is their capacity to sustain high transport currents. Under isothermal conditions ($T < T_c$), for each value of the induction in the range $B_{c1} < B < B_{c2}$ there is a specific maximum value of the current density J_c which if exceeded causes the superconductor to pass into the normal state. The current density $J_c(B, T)$ is called the critical value for specified values of B and T.

The character of the $J_c(B)$ relationship for nonideal superconductors of the second kind is illustrated in Fig. 2-4 as a set of isotherms relating to Nb–Zr and Nb–Ti alloys.

It is quite clear from the foregoing that, whereas for ideal superconductors of the second kind, the geometrical locus of the points corresponding to the transition from the mixed ($B_{c1} < B < B_{c2}$) to the normal ($B = B_{c2}$) state forms a curve in the B–T coordinate system, for nonideal superconductors of the second

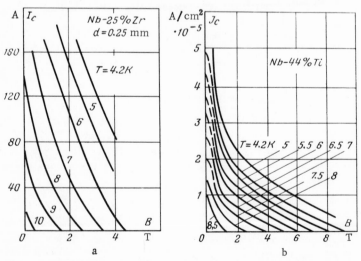

Fig. 2-4. Critical current density for nonideal superconductors of the second kind.

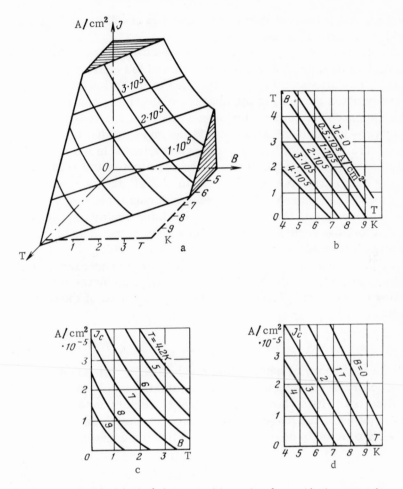

Fig. 2-5. Geometrical loci of phase-transition points for nonideal superconductors of the second kind.

kind this geometrical locus of the points forms a surface in the $B-T-J$ system. By way of example, Fig. 2-5 shows a surface based on experimental data obtained for heat-treated $Nb-25\%Zr$ alloy wire [25]. Projections of this surface on the coordinate planes respectively give the $B-T$, $J-B$, and $J-T$ diagrams (Fig. 2-5b-d). The $B-T$ diagram shows the temperature dependence of the second critical induction for various values of the current density. We see from this diagram, in particular, that

the higher the current density the lower is the critical magnetic
field at a particular temperature.

The capacity of nonideal superconductors of the second kind
to pass substantial transport currents and the character of the
$J_c(B, T)$ relationship may be explained in the following way. As
already noted earlier, in the absence of detaining centers (i.e., in
ideal superconductors of the second kind) the vortices form a regular two-dimensional structure in the superconducting matrix.
Repulsive forces act between any two vortices. In the case of a
regular structure of the vortices the resultant of these forces F_r
acting on a particular vortex by virtue of all the surrounding
vortices is, of course, equal to zero. For an irregular structure
(such as occurs in the presence of detaining centers in the superconductor) the resultant force differs from zero. The force constitutes a peculiar kind of pressure "from outside," i.e., from the
direction in which the external magnetic field penetrates into the
superconductor. Clearly this force is balanced by the force F_{det}
acting on the particular vortex from the direction of the detaining
center

$$F_r = F_{det}. \qquad (2-16)$$

In the theory of equilibrium magnetization for nonideal superconductors of the second kind based on this model (Silcox and
Rollins [26]), it is shown that, subject to certain assumptions, the
distribution of the magnetic induction over the cross section of a
superconductor is described by the following equation:

$$\frac{\partial B^2}{\partial x} = \text{const}, \qquad (2-17)$$

where x is the distance from the surface of the superconductor.
The distribution of induction over the cross section of a plate made
from a nonideal superconductor of the second kind in a field of
induction $B > B_{c1}$ is shown schematically in Fig. 2-6a (continuous
line). Figure 2-6b shows the distribution of induction over the
cross section of an ideal superconductor of the second kind. In
this case, as a result of the absence of detaining centers, the
induction is distributed uniformly.

On passing a current through the superconductor which is not
in the same direction as the magnetic field, a Lorentz force F_L

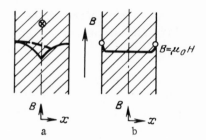

Fig. 2-6. Distribution of the induction over the cross sections of nonideal and ideal superconductors of the second kind.

acts on the vortices. In this case the resultant force is added to the Lorentz force and their sum is balanced by a force acting by virtue of the det?ining center, i.e., for $J > 0$

$$F_r + F_L = F_{det}.$$ (2-18)

The limiting state in which equilibrium may just be achieved between these forces corresponds to a "smoothed" distribution of induction, with a certain maximum slope at each point of the sample, i.e., with a certain maximum current density (the so-called critical state). This maximum current density, generally speaking, depends on the induction at the particular point of the sample, and hence the lines representing the distribution of the induction over the sample cross section possess a certain curvature (Fig. 2-6a).

It should be remembered that, although in the critical state, there is a certain maximum current density at every point of the sample the sample may not carry any transport current when considered as a whole, since the current density may be oriented in opposite directions at different points of the cross section, as indicated, for example, in Fig. 2-6a. Reaching the critical current of the sample means that the current density is directed in the same way at any point of the cross section.

On increasing the current density above a certain maximum value the Lorentz force becomes greater than the detaining force, and the vortex starts moving. As already mentioned, this leads to the continuous evolution of Joule heat, the temperature may rise above T_c, and the superconductor will then pass into the normal state.

It follows from the foregoing that for specified values of the induction B and ambient temperature T the critical current density is that for which $F_L = F_{det}$. Thus

$$F_L = F_{det} \text{ for } J = J_c; \qquad (2\text{-}19)$$

$$F_L > F_{det} \text{ for } J > J_c. \qquad (2\text{-}20)$$

The reason for the transition of the nonideal superconductor of the second kind into the normal state on exceeding the critical current in a steady external magnetic field is ultimately the same as that of the transition of an ideal superconductor of the first kind; it is due to exceeding the critical temperature T_{cr} (B) of the superconductor for the particular value of B.

Clearly, the greater the Lorentz force which the detaining centers are able to "hold back," the greater is the current capable of being passed through a nonideal superconductor of the second kind. The vortices move from one detaining center to another under the influence of the Lorentz force (moving in the direction of the latter). In order to overcome the detaining center the vortex has to perform work equal to $F_L d$, where d is the size of the detaining center. In other words, the vortex overcomes a certain potential barrier.

It follows from Eq. (2-14) that the Lorentz force acting on the vortex is proportional to the product

$$a = JB. \qquad (2\text{-}21)$$

In the case of the critical current density

$$a_c = J_c B. \qquad (2\text{-}22)$$

It is clear that if $JB < \alpha_c$ the system will be stable. If $JB > \alpha_c$ there will be a continuous motion of the vortices and the superconductor will pass into the normal state.

The value of α_c (called the Lorentz force criterion) depends rather heavily on the degree of structural disorder in the material.

It should be noted that, in actual fact, according to experimental data [27-29], the relationship between the quantities B and J differs slightly from (2-22); it is found that

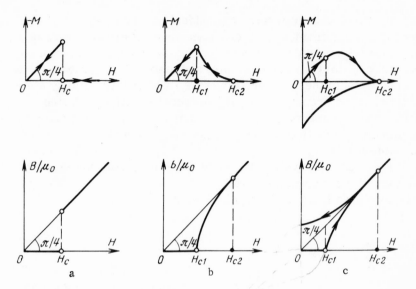

Fig. 2-7. The M—H and B—H diagrams. a) Ideal superconductor of the first kind; b) ideal superconductor of the second kind; c) nonideal superconductor of the second kind.

$$J_c(B+B_0) = \alpha_c, \qquad (2\text{-}23)$$

where B_0 is a constant. However, since the induction B_0 is relatively weak (for the alloy Nb—25% Zr, $B_0 = 0.1$ to 0.2 T, for Nb_3Sn a few tenths of a tesla), in the case of strong fields we may neglect B_0 by comparison with B in (2-23).

Thus α_c is proportional to the force F_{det} acting on the vortex from the direction of the detaining center. In experiments by Kim et al. [27-29] it was found that the values α_c and hence F_{det} were related linearly with temperature.

It was also found that each detaining center was associated, not with individual vortices, but with groups of vortices having a size of the order of λ, i.e., groups incorporating several tens of fluxoids. In particular, on magnetizing nonideal superconductors of the second kind with a monotonically increasing magnetic induction (in the absence of transport currents) the penetration of the field into the sample took place in discrete "steps," the height of these steps corresponding to the magnetic flux carried by one group of fluxoids.

A characteristic feature of nonideal superconductors of the second kind distinguishing them from ideal superconductors is the irreversible character of the transition from the superconducting (mixed) into the normal state, and also that of the magnetization process in these superconductors. The irreversibility of the magnetization process is due to the fact that both increasing and decreasing the magnetic field in the sample involve doing work in moving the vortices to which the forces from the detaining centers are applied.

Figure 2-7 shows the M−H and B−H diagrams for superconductors of the types considered.

It follows from the general equation

$$B = \mu_0(H+M), \qquad (2\text{-}24)$$

where

$$M = \chi H, \qquad (2\text{-}25)$$

that for superconductors exhibiting the Meissner effect, i.e., superconductors of the first kind with $B < B_c$ and superconductors of the second kind with $B < B_{c1}$,

$$B_s = 0. \qquad (2\text{-}26)$$

The magnetic susceptibility χ is determined by the obvious equation

$$\chi_s = -1, \qquad (2\text{-}27)$$

and hence the magnetization of the superconductor is given by

$$M_s = -H. \qquad (2\text{-}28)$$

Since as already indicated a superconductor has hardly any magnetization in the normal state, for this state (subject to the condition $B > B_c$ in superconductors of the first and $B > B_{c2}$ in superconductors of the second kind) we shall have

$$B_n = \mu_0 H \qquad (2\text{-}29)$$

and

$$M_n = 0. \qquad\qquad (2\text{-}30)$$

We see from the diagrams presented in Fig. 2-7a, b that the position of the magnetization curves for ideal superconductors of the first and second kinds is naturally independent of whether the external magnetic field is rising or falling along these curves. For nonideal superconductors of the second kind the position of these curves depends very considerably on the direction in which the field strength is changing. A characteristic hysteresis loop appears on the magnetization curves, and on reducing the magnetic field to zero, residual magnetization appears in superconductors of this type (Fig. 2-7c). The form of these curves depends on the degree of disorder in the crystal structure, i.e., on the number of detaining centers and on the threshold energy of their attachment to the vortices.

The critical current density depends very much on the composition of the superconductor (in the case of alloys), the degree of cold deformation (working) of the alloy (the higher the degree of deformation, the greater is the number of dislocations and other defects inside the material), the heat treatment applied, and various other factors. Annealing, in particular, may lead to the decomposition of the high-temperature phase, as a result of which stresses appear in the crystal lattice.

2.3. Creep and Jumps in Magnetic Flux in Nonideal Superconductors of the Second Kind

One of the most important effects encountered in nonideal superconductors of the second kind is that of "jumps" in the magnetic flux and the flux "creep" which produces them.

As we have already discussed, the vortices in a nonideal superconductor of the second kind move continuously for $J > J_c$ (for $B > B_{c1}$ when $T < T_c$). When the current density reaches the critical value ($J = J_c$) on increasing the current in the superconductor, the distribution of the magnetic induction over the cross section of the conductor becomes monotonic, but as yet there is no continuous motion of the vortices; this only begins on exceeding the critical current density. It is nevertheless found that individual groups of vortices may "break away" from one detaining center and pass to another even for a current density of $J < J_c$.

This motion of the vortex groups is of a fluctuational nature. Let us consider a certain detaining center with a group of vortices attached to it. If as a result of fluctuations there is a local temperature rise in the neighborhood of the detaining center in question, F_{det} will diminish. The reduction in F_{det} is due to the fact that the Lorentz force criterion α_c falls with increasing temperature, i.e., the threshold binding energy between the group of vortices and the detaining center is reduced. The value of F_{det} may be smaller than the Lorentz force F_L acting on the group of vortices under consideration; this group will then break away and move to the next detaining center. This fluctuational motion of the groups is called flux creep. Clearly as a result of creep the distribution of induction in a nonideal superconductor of the second kind will become more and more uniform over the cross section as time progresses. From this point of view flux creep may be regarded as a special kind of "diffusion" of the magnetic flux inside the sample.

During the motion of a group of vortices between detaining centers an emf directed along the transport current acts in the system and a certain amount of Joule heat is developed. This heat evolution raises the temperature in this particular region of the superconductor. Under certain conditions, which we shall subsequently indicate, this rise in temperature may be so considerable that it will lead to a substantial reduction in the F_{det} of the neighboring detaining centers and hence to the detachment of groups of vortices from these also. As each of these groups of vortices moves there will be a further evolution of Joule heat, the surrounding parts of the superconductor will become heated, and so on. This will give rise to an avalanche process leading to a major redistribution (partial equalizing) of the magnetic induction in the superconductor. This process is called flux jumping. According to the estimates of Anderson and Kim [30] a flux jump develops when the local rise in temperature amounts to at least 0.01–0.001 of the actual sample temperature.

The total amount of Joule heat evolved may be so great that it will lead to a major rise in the temperature of a particular region of the superconductor, much greater than the original fluctuational temperature increment of the detaining center. If the temperature then exceeds the critical temperature for the particular magnetic field $T_c(B)$ a normal zone will develop in the superconductor, and under certain conditions (e.g., inefficient removal of the Joule heat evolved in this normal zone) this may extend further and convert the whole superconducting sample into the normal state. Such a flux jump (causing the whole superconductor to return to the normal state) is sometimes called a "catastrophic" jump. As a result of the "catastrophic" flux jump the distribution of the induction will become uniform over the whole sample cross section (as in the case of an ideal superconductor of the second kind, or on reaching the critical current density in a nonideal superconductor of the second kind).

Thus, subject to the conditions $B_{c1} < B < B_{c2}$ and initial superconductor temperature $T < T_c(B)$, a nonideal superconductor of the second kind carrying a transport current may pass into the normal state, not only on exceeding the critical current density $I_c(B, T)$, but also when $J < J_c$, by virtue of a flux jump.

Flux creep by no means always leads to the development of jumps. Flux jumps are most likely to occur under specific favorable circumstances. Let us consider for what values of the external magnetic field and for what temperature of the superconductor jumps are most likely to arise.

We see from Eqs. (2-14) and (2-21) that the smaller the value of B, the greater will the current density J have to be in order to give the same value of the Lorentz force. From the point of view of the possible breakaway of a group of vortices from a particular detaining center, it is important to know the absolute value of F_L, and it is immaterial for what values of J and B this is achieved. It is also quite obvious that, during the motion of a group after its detachment from the detaining center, the Joule heat release will be so much the greater, and hence the rise in the temperature of neighboring centers so much the more substantial, the higher the current density J. It follows that from the point of view of the probability of the development of a flux jump the most dangerous region is that characterized by low values of the external magnetic field, for which the current density in the superconductor may be quite high.

As regards the temperature of the superconductor, the most dangerous region is the low-temperatures region (below the boil-

ing point of helium at atmospheric pressure), since with falling
temperature there is a sharp reduction in the thermal conductivity
of the superconductor, and its specific heat also diminishes, so
that at low temperatures the conditions of heat release and heat
accumulation worsen. Hence the same amount of Joule heat evolved
during the displacement of groups of vortices during the creep process will lead to a much greater local temperature rise at low than
at higher temperatures.

Finally it should be emphasized that by no means will every flux
jump lead to the formation of a normal zone in the superconductor. The question as to the precise conditions under which
a jump will become "catastrophic" may be considered from the
purely thermodynamic point of view [31]. Since a flux jump occurs
in a very short period of time, the redistribution of the magnetic
flux over the cross section of the superconducting sample may be
regarded as a practically adiabatic process (one in which the entropy of the sample remains constant). This process takes place
for a constant external magnetic field (and constant pressure p).

It is well known [14] that for a system existing in a magnetic
field (duly matched with the ambient; H = const, p = const, and
S = const), the condition of thermodynamic equilibrium may be
written in the form

$$di^* \leqslant 0, \tag{2-31}$$

where i^* is the enthalpy of the magnetic material defined by the
equation

$$i^* = i - HM. \tag{2-32}$$

Here i is the enthalpy of the magnetic material in the absence of a magnetic field, p is the pressure, and S is the entropy. The
equality sign in Eq. (2-31) corresponds to the state of equilibrium
of the system and the inequality sign to the original, nonequili-
brium state, in which the system is "on the way" to a state of
equilibrium.

Figure 2-8a shows the enthalpy as a function of the magnetic
field induction B for a uniform distribution of the magnetic induction over the cross section of a nonideal superconductor of the
second kind (broken line). As already mentioned, the distribution

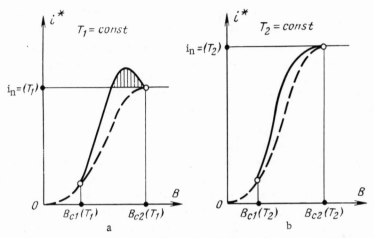

Fig. 2-8. Enthalpy of a superconductor in relation to the magnetic induction.

of the induction has the same character in this case as in an ideal superconductor of the second kind. It follows from the well-known thermodynamic relation [14]

$$\left(\frac{di^*}{dH}\right)_{T,\,p} = -M \tag{2-33}$$

that

$$i^*(H,\ T) = i\,(0,\ T) - \int_0^H M\,dH. \tag{2-34}$$

The broken curve in Fig. 2-8a is calculated from this equation on the basis of the well-known dependence of the magnetization M of an ideal superconductor of the second kind on the field H (Fig. 2-7b). *

Figures 2-8a,b also show the $i^* = f$ (B) relationship for a nonideal superconductor of the second kind in the case of a non-uniform distribution of the induction over the cross section for two different isotherms T_1 and T_2 ($T_1 < T_2$). This relationship is calculated with the aid of Eq. (2-34) on the basis of existing data regarding the magnetization of a nonideal superconductor of the second kind (Fig. 2-7).

---

* Clearly for B $\leq B_{c1}$ the $i^* = f$(H) relationship has the character of a quadratic parabola [14].

We see from Fig. 2-8 that for $B_{c1} < B < B_{c2}$ the enthalpy of a superconductor with a nonuniform distribution of induction is higher than that of the same superconductor with a uniform distribution. Hence in accordance with Eq. (2-31) spontaneous processes in the system will tend to establish a uniform distribution of the magnetic induction, i.e., flux creep is inevitable.

The quantity i_n in Fig. 2-8 denotes the enthalpy of the superconductor in the normal state. We see from this figure that at lower temperatures the enthalpy of a superconductor in the state of nonuniform magnetization is greater than the enthalpy in the normal state. Hence, for a particular sample of nonideal superconductor of the second kind existing at a temperature T_1 in the range of external magnetic field values corresponding to the shaded region (Fig. 2-8), spontaneous processes in the system under consideration will, in conformity with Eq. (2-31), tend to produce a transition into the normal state; i.e., a "catastrophic" flux jump may occur.

In this case the "catastrophic" jump will only occur when the conditions of heat release from the sample into the helium tank are poor and the magnetic induction becomes equalized in an avalanche manner over the whole sample cross section. In the case of good heat release the avalanche will not develop and no "catastrophic" flux jump will occur.

Thus for those cases in which the $i^* = f(B)$ relationship has the form of Fig. 2-8b the transition of the system into the normal state as a result of a flux jump is in principle impossible, i.e., in this case the avalanche-like equalization of the magnetic induction over the whole sample cross section will not lead to the sample being heated above the critical temperature $T_c(B)$. However, in cases analogous to those shown in Fig. 2-8a, a transition into the normal state as a result of a "catastrophic" jump may in fact take place. The character of the curves shown in Fig. 2-8 depends not only on the sample temperature but also on its size. The greater the cross section of the sample, the greater is the difference between the induction at the surface and in the center (this may readily be seen, for example, from Fig. 2-6a), and ultimately the greater will be the difference between the enthalpy of the sample in the initial state and in the state having a uniform distribution of the induction over the cross section. It may happen that for one particular sample temperature and a small sample diameter the continuous curve in Fig. 2-8 will everywhere lie below i_n, while for a sample of large diameter this curve will lie above i_n over a certain range of field values. Thus in this range a "catastrophic" flux jump may take place.

We may arrive at the same conclusions on the basis of other arguments. We may consider, for example, that the "intensity" of a jump is proportional to the sample cross section. A jump having an "intensity" over a certain limit may prove to be "catastrophic." It is clear from the foregoing that this limit will be the smaller, the lower the sample temperature.

It follows from Fig. 2-8 that, as i_n is greater than the enthalpy of the superconductor with a uniform distribution of magnetic induction for any $B < B_c$, the state with the uniform distribution is thermodynamically more advantageous than the normal state (for specified T and B values); this conclusion is trivial. Under equilibrium conditions, when no external factors promoting the growth of the normal zone are operative, the sample will pass into the superconducting state, with a uniform distribution of magnetic induction over the whole sample cross section.

2.4. Resistive State of Nonideal
Superconductors of the Second Kind

It was noted earlier that, when the current in a nonideal superconductor of the second kind exceeded the critical value (for specified B and T values), continuous motion of the vortices occurred, and hence so did a continuous evolution of Joule heat. As a result of this, the superconductor is heated above the critical temperature (for the specified field) and passes into the normal state.

In addition to this, however, we may encounter a situation in which the transfer of the evolving Joule heat to the ambient (liquid helium) is organized so intensively that dynamic equilibrium between the evolving and outgoing heat is established at a superconductor temperature above the temperature of the helium tank T_t but lower than the critical temperature in the specified field $T_c(B)$, i.e.,

$$T_t < T < T_c(B). \qquad (2-35)$$

This situation is of special interest from the point of view of

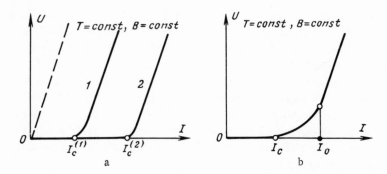

Fig. 2-9. Volt—ampere characteristics for a nonideal superconductor of the second kind.

thermal-stabilization problems; it was studied in great detail in experiments by Kim and his colleagues [32, 33].

It was found that if a current $I > I_c$ was passed through a nonideal superconductor of the second kind a potential difference U, increasing almost linearly with rising current, was established between the ends of the sample. The volt — ampere characteristics for this case are illustrated schematically in Fig. 2-9. These characteristics relate to isothermal conditions for $T < T_c(B)$ and for a constant magnetic induction.

Figure 2-9a shows the volt—ampere characteristics of two samples of the same nonideal superconductor of the second kind but with different values of the critical current, i.e., different numbers of structural defects. This difference may be associated with the conditions of mechanical processing and heat treatment.

For small values of $(I - I_c)$ the U(I) relationship is nonlinear; with increasing $(I - I_c)$ linearity sets in. The initial part of the U(I) curve is shown on a large scale in Fig. 2-9b; it should be noted that for $I = I_c$ the U(I) curve joins the horizontal axis tangentially.

The broken line in Fig. 2-9a indicates the hypothetical volt — ampere characteristic for an ideal superconductor of the second kind in the case $I_c = 0$.

Although the lines U(I) start from different points on the horizontal axis for different I_c, they remain equidistant (for the same T).

Figure 2-10 [32] shows some experimental U(I) relationships for a Nb—50%Ta alloy in the resistive state (Fig. 2-10a — the curves relate to samples with different numbers of defects); similar curves for Pb−17% In appear in Fig. 2-10b. These data refer to isothermal conditions. Of course, if the helium tank is kept at a constant temperature T_t, it is fundamentally impossible to ensure precise isothermal conditions for different currents I, since the amount of Joule heat liberated increases with I (and hence $I - I_c$), and the temperature at which thermal equilibrium of the superconductor is attained does likewise. In order to obtain data consistent with isothermal conditions it is thus essential to analyze the results of a number of experiments executed for different temperatures of the helium tank.

The properties of a nonideal superconducting sample of the second kind (neglecting the difference $I_c - I_0$) may under these circumstances be characterized (Fig. 2-9b) by a resistance value as follows:

$$R_{\text{res}} = \frac{U}{I - I_c}. \tag{2-36}$$

It follows from the foregoing that at a specified temperature R_{res} will not depend on I for $I > I_c$ and will be independent of I_c, i.e., it will be the same for samples of the same material having different values of I_c. In the current range I_c to I_0 the sample resistance increases monotonically from zero to R_{res} (Fig. 2-9b). Clearly R_{res} changes from zero in the superconducting state (for $I < I_c$) to the resistance of the superconductor in the normal state R_n for $T > T_c(B)$ or $B > B_{c2}(T)$.

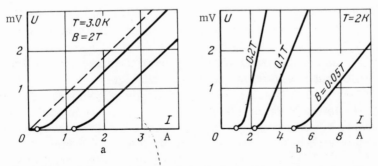

Fig. 2-10. Volt−ampere characteristics in the resistive state. a) For alloy Nb−50% Ta (samples containing different numbers of defects); b) alloy Pb−17% In.

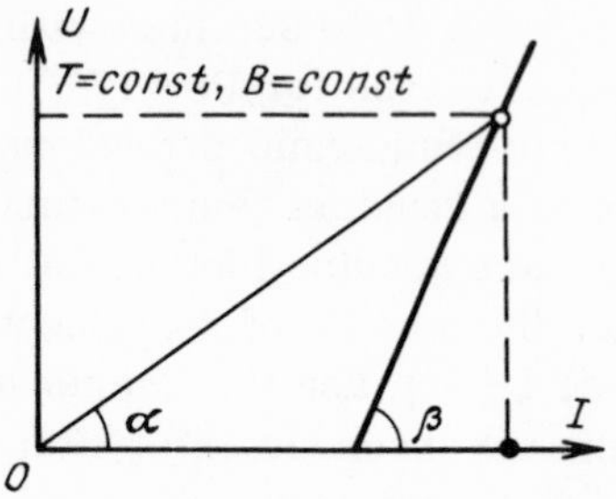

Fig. 2-11. Volt–ampere characteristic of a superconductor in the resistive state R_{res} = tan β ; R_{res}^{qu} = tan α.

Thus, under conditions of intense heat exchange with the helium tank at $I > I_c$ the sample passes into a special physical state which is not superconducting (since $R_{res} > 0$) but cannot be regarded as the normal state (since $R_{res} < R_n$). This state of non-ideal superconductors of the second kind is called the resistive state.

Thus for a superconductor in the resistive state Ohm's law is not satisfied; there is no proportionality between U and I. Let us introduce the concept of the quasi-ohmic resistance of a superconductor

$$R_{res}^{qu} = \frac{U}{I}. \tag{2-37}$$

Allowing for (2-36) we thus obtain the following equation:

$$R_{res}^{qu} = R_{res}\frac{I - I_c}{I}. \tag{2-38}$$

The meaning of R_{res} and R_{res}^{qu} is illustrated by Fig. 2-11.

Experimental data characterizing the resistive state are extremely sparse. This is because of the great complexities involved in such experiments, arising from the intensive outflow of the Joule heat liberated in the superconductor.

Figure 2-12a illustrates the resistance of a superconductor in the resistive state as a function of the magnetic induction for I = const, T = const, and $I > I_c$. Of course for T = const each B value has its own I_{cr}. For B = 0 we have $R_{res} = 0$, since in the absence of a field there are no vortices in the superconductor. For B = $B_{c2}(T, J)$ the superconductor passes into the normal state and R_{res} becomes equal to R_n. Figure 2-12b shows the experi-

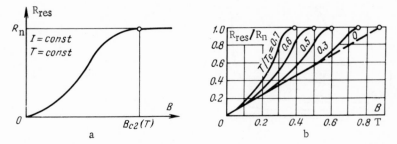

Fig. 2-12. Resistance of a superconductor in the resistive state as a function of the external magnetic field.

mental dependence of R_{res}/R_n on B for a sample of Nb−50%Ta alloy at various temperatures [33]. The dimensionless values of the temperature T/T_{c0} are shown on the isotherms, in which T_{c0} is the critical temperature of the superconductor for B = 0. For the alloy Nb−50%Ta T_{c0} = 6.15 K, $B_{c2}(0)$ = 0.86 T.

It was shown by Strnad et al. [32] that the relationship expressed in reduced parameters

$$R_{res}/R_n = f\left(\frac{T}{T_{c0}}, \frac{B}{B_{c2}}\right) \tag{2-39}$$

obeyed the law of corresponding states, i.e., the values of R_{res}/R_n were the same for different nonideal superconductors of the second kind if the values of $\tau = T/T_{c0}$ and $B/B_{c2}(T)$ were kept constant. This law holds for superconductors having not too high a value of the Ginzburg−Landau parameter. As $T/T_{c0} \to 0$, Eq. (2-39) acquires an extremely simple form:

$$R_{res}/R_n = \frac{B}{B_{c2}(0)}, \tag{2-40}$$

which corresponds to the broken line in Fig. 2-12b.

Equation (2-39) may be written in the following way:

$$R_{res}/R_n = \frac{B}{B_{c2}(T)\, g(\tau)}. \tag{2-41}$$

If we write

$$B_{c2}(T) = B_{c2}(0)\, h(\tau), \tag{2-42}$$

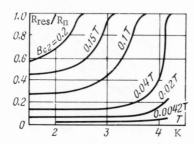

Fig. 2-13. Temperature dependence of the resistance of a superconductor in the resistive state.

Eq. (2-41) may be expressed in the form

$$R_{res}/R_n = \frac{B}{B_{c2}(0)\,\varphi(\tau)},$$ (2-43)

where

$$\varphi(\tau) = g(\tau)h(\tau).$$ (2-44)

Clearly for $\tau = 0$ we obtain $\varphi = 1$; with rising temperature the value of $\varphi(\tau)$ diminishes. We see from Fig. 2-12b that over a particular range of τ values (depending on B) the value of R_{res}/R_n remains independent of temperature, i.e., $\varphi(\tau) = 1$; thus for B $\leq$ 0.5 T this ratio is temperature-independent over the range of τ values between 0 and approximately 0.3. On raising the temperature the value of R_{res}/R_n increases.

The character of the temperature dependence of the resistance of a nonideal superconductor of the second kind in the resistive state is illustrated in Fig. 2-13, which shows the $R_{res}/R_n = f(T)$ relationship for various values of B_{c2} in the case of Nb$-$50%Ta [32]. The isotherms on the volt$-$ampere characteristic are illustrated in Fig. 2-14. The higher the temperature, the greater is

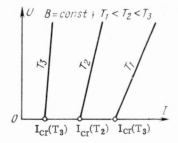

Fig. 2-14. Isotherms in the volt$-$ampere characteristic of a nonideal superconductor of the second kind.

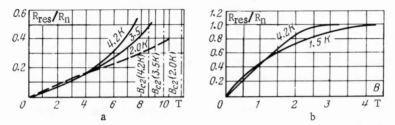

Fig. 2-15. Resistance of a superconductor in the resistive state as a function of the external magnetic induction. a) For a Ti−50% V alloy; b) for a Nb−10% Ti alloy.

the value of R_{res} and hence the greater is the slope of the isotherm. Since I_{cr} diminishes with rising temperature, we find that the higher the temperature T the closer do the isotherms move toward the origin of coordinates.

The results presented refer to nonideal superconductors of the second kind (Nb−Ta and Pb−In) with fairly low values of the Ginzburg−Landau parameter $\varkappa < 5$ and low critical fields ($B_{c2} < 1$ T). The heavy-current superconductors with high values of B_{c2} of greatest practical interest, such as Nb−Zr, Nb−Ti, V−Ti, and other alloys, for which $\varkappa \approx 10 - 100$, are characterized by B and T dependence of R_{res}/R_n having a completely different character.

Figure 2-15a shows the dependence of R_{res}/R_n on B at various temperatures for a Ti−50% V alloy and Fig. 2-15b the same for Nb−10% Ti; Fig. 2-16 shows the temperature dependence of R_{res}/R_n for various values of B and the same alloy. Unfortunately

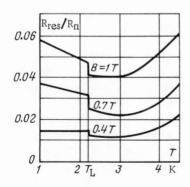

Fig. 2-16. Temperature dependence of R_{res}/R_n for a Nb−10% Ti alloy.

there are no data regarding the temperature dependence of R_{res}/R_n for superconductors of this type in the temperature range $T > 4.2$ K, which is of greatest practical interest. As indicated in Chapter 5, this lack of data impedes consideration of the processes taking place in combined superconductors.

Clearly the intensity of heat evolution associated with the transport current I through a semiconductor in the resistive state will be given by the equation

$$W = UI. \tag{2-45}$$

It follows from (2-36) that

$$U_{res} = R_{res}(I - I_c), \tag{2-46}$$

so that (2-45) may be expressed in the form

$$W = R_{res} I (I - I_c), \tag{2-47}$$

or, on allowing for (2-38), in the form

$$W = I^2 R_{res}^{qu}. \tag{2-48}$$

Allowing for (2-34) we obtain

$$W = \frac{B R_n}{B_{c2}(0)\, \varphi\,(T)}\, I\,(I - I_c). \tag{2-49}$$

It is clear from Eq. (2-49) that the intensity of heat evolution remains constant in time.

This result is particularly interesting. It was in fact noted earlier that for $I > I_c$ a force $F_L - F_{det}$ acted upon the vortex. It would appear that under the influence of this force the vortex should accelerate, and that the intensity of Joule heat evolution should therefore increase continuously during the motion of the vortex. The foregoing experimental results may be explained if we assume that the ambient offers resistance to the motion of the vortices. If we consider that this resistance has a quasi-viscous nature and hence increases with increasing velocity of the vortex, for a specific velocity v the force $F_L - F_{det}$ will be balanced by the resistance of the ambient

$$F_L - F_{det} = \eta v, \tag{2-50}$$

where η is a constant which may be regarded as a viscosity coefficient.

The Lorentz force acting on unit length of the vortex will be given by the equation

$$F_L = \varphi_0 l. \tag{2-51}$$

Since when $I = I_c$, $F_L = F_{det}$, the force acting on the vortex from the detaining center may be determined in the following way from (2-51):

$$F_{det} = \varphi_0 I_c. \tag{2-52}$$

Using (2-50)-(2-52), we obtain the following relationship for the velocity of the vortex:

$$v = \frac{\varphi_0}{\eta}(I - I_c). \tag{2-53}$$

It is easy to see that the intensity of heat evolution per unit volume of the superconductor during the motion of the vortices will be given by the equation

$$W_V = n F_L v, \tag{2-54}$$

where n is the number of vortices per unit cross section of the superconductor. Allowing for (2-53) we then have

$$W_V = \frac{n\varphi_0^2}{\eta} I (I - I_c). \tag{2-55}$$

Since

$$B = n\varphi_0, \tag{2-56}$$

while for a thin sample we may consider that

$$B \approx \mu_0 H, \tag{2-57}$$

Eq. (2-55) may be written in the following form

$$W_V = \frac{\varphi_0 B}{\eta} I (I - I_c).$$ (2-58)

Equation (2-58) coincides with (2-49) if we consider that

$$\eta = \frac{\varphi_0 B_{c2} (0) \varphi (T)}{\rho_n}.$$ (2-59)

In this equation the resistivity ρ_n enters instead of the absolute resistance R_n since we have transformed from the total intensity of heat evolution W to the heat evolution per unit volume W_V in (2-49). We see from Eq. (2-59) that the viscosity η is constant for a specified temperature.

It is important to note that if isothermal conditions are ensured for a superconductor in the resistive state then in principle any arbitrarily heavy current may be passed through it. In actual practice only the temperature of the ambient (the helium tank) can be guaranteed to be constant (T_t = const).

Under conditions of thermal equilibrium the intensity of heat evolution in the resistive impedance W is equal to the intensity of heat outflow from the surface of the superconductor into the helium tank

$$W = hPl(T - T_t),$$ (2-60)

where h is the heat-transfer coefficient to the liquid helium (a function of temperature); P and l are respectively the perimeter and length of the sample.

Allowing for (2-47)

$$R_{res} l (I - I_c) = hPl(T - T_t),$$ (2-61)

whence

$$T = T_t + \frac{R_{res}}{hPl} I (I - I_c).$$ (2-62)

Thus the higher the value of I, the higher is the temperature of the superconductor. Of course for a certain current the temperature of the superconductor may exceed the critical temperature $T_c(B)$ and the superconductor will pass into the normal state.

Chapter 3

Protection of Superconducting
Magnetic Systems

3.1. General Principles

As already mentioned, the energy of the magnetic field capable of being stored in more or less large superconducting magnetic systems may rise to an extremely high value. Thus in the superconducting magnetic system of the bubble chamber of the Argonne National Laboratory (USA) the stored energy amounts to 80 MJ; in the magnetic system of the bubble chamber in the National Accelerator Laboratory of the USA the energy reaches 396 MJ.

In a number of cases superconducting magnetic systems are made specially with the intention of storing energy. Naturally the density of the stored energy increases with increasing magnetic field. For an induction of B = 15 T, the maximum which has so far been achieved, the energy density is 90 kJ/liter [34].

If there is an accidental transition of the superconducting magnetic system into the normal state, the stored energy is released in the form of Joule heat in that part of the winding which has become normal. This heat evolution may heat neighboring parts of the winding, and these regions will also pass into the normal state.

Under certain conditions the propagation of the normal zone through the winding acquires an avalanche-like character. Subsequently we shall call this uncontrollable process an unstabilized transition of the winding into the normal state.

The accidental (uncontrollable) transition of the superconducting magnetic system into the normal state is accompanied by a number of undesirable effects.

45

Firstly, the evolution of a large amount of Joule heat in a relatively small part of the winding (that which has passed into the normal state) may break (or melt) that part of the winding.

Secondly, the large potential difference developing during the transition at the ends of the section which has passed into the normal state may lead to the breakdown of the insulation between the turns and to the short-circuiting of a section of the coil. Clearly on subsequent excitation of the solenoid these short-circuited turns will play the part of a secondary transformer winding. A large current will be induced in these turns, and this will lead to the premature transition of the coil into the normal state (i.e., for low values of the current fed into the winding from the external source).

A good idea of the principal characteristics of the transition of a superconducting system into the normal state may be obtained, for example, from Fig. 3-1. This figure illustrates the experimental values of the coil current i, the voltage at the ends of the section of winding which has passed into the normal state (u), the resistance of this section (R), and the instantaneous power of the heat evolution (w) for one of the smaller experimental solenoids. The internal diameter of the solenoid is 16 mm, the external diameter 48.5 mm, the height 54.5 mm, the number of turns 7600, the inductance 0.735 H; the winding material is a copper-free wire 0.25 mm in diameter composed of the ternary alloy 65BT (niobium 65 at.%, titanium 25 at.%, balance zirconium and technological additives [35]). The data presented in Fig. 3-1 relate to a transient process for an initial current of 10 A in the solenoid (the critical current for this solenoid is 12 A). For this current the energy stored in the solenoid is 37 J. We see from the figure that the voltage at the ends of the normal section reaches U_{max} = 1100 V, while the instantaneous power of heat evolution is W_{max} = 7.25 kW. The transient process here takes place in an extremely short time; for example, it reaches its maxi-

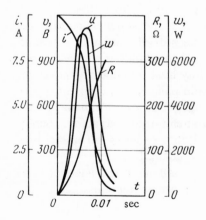

Fig. 3-1. Characteristics of the transition of a superconducting system into the normal state.

mum in 0.0065 sec. For the critical current of 12 A, at which the stored energy is 53 J, U_{max} reaches 2000 V, while W_{max} exceeds 14 kW. We see from these figures that even for a small solenoid the power of heat evolution concentrated in a comparatively small part of the winding and the overvoltage at the ends of the normal section reach large values. In larger solenoids with a stored energy of the order of tens of kilojoules, these values become very much greater.

Of course for such systems an uncontrolled passage into the normal state may break the winding. Superconducting systems with large stored energies are therefore not usually set up without taking a whole series of measures to prevent an unstabilized transition into the normal state.

Thirdly, the Joule heat liberated during the transition of the system into the normal state is taken up by the liquid helium filling the cryostat of the magnetic system. It is well known that liquid helium has an extremely low heat of vaporization r; at atmospheric pressure r = 20.6 kJ/kg = 2.68 kJ/liter.* Hence the assimilation of 1 kJ by liquid helium leads to the evaporation of 0.374 liter of helium. Thus as a result of the liberation of the Joule heat in the cryostat during the transition a great deal of helium evaporates. If provision is not made for the release of all the evaporating helium from the cryostat, the pressure in the latter will start rising, and since the heat of vaporization diminishes with increasing pressure (Fig. 3-2) for the same rate of heat liberation the intensity of helium evaporation will increase still further. If provision is not made for the release of the vapor, the unintended rise in cryostat pressure may lead to an explosion. It should also be remembered that for a certain critical decrease

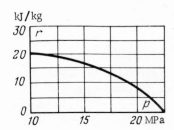

Fig. 3-2. Heat of vaporization as a
function of pressure for helium.

---

*We note for comparison that the heat of vaporization of water at atmospheric pressure is 2256.7 kJ/kg = 2162.8 kJ/liter.

in pressure between the cryostat and the receiving vessel the velocity of the helium vapor flowing out of the cryostat will reach the local velocity of sound in helium vapor, and the outlet pipe will be blocked. Any further increase in pressure in the cryostat will not raise the rate of flow of the vapor through the pipeline [36]. Hence the pressure in the cryostat will increase sharply, and this also may lead to an explosion.

Thus the uncontrolled transition of a superconducting magnetic system into the normal state constitutes a serious emergency which may lead to the partial or complete destruction of costly equipment. It is accordingly most vital to develop efficient methods of protecting superconducting magnetic systems to counteract an accidental transition into the normal state. If the transition of a superconducting coil into the normal state cannot actually be prevented, the protective system should pursue the following three principal aims:

1. To reduce the overvoltage in the section of winding which has passed into the normal state in order to prevent the breakdown of the insulation between the windings.
2. To reduce the proportion of stored energy liberated in the section passing into the normal state in order to prevent the burning-out of the winding.
3. To reduce the proportion of stored energy liberated inside the cryostat in order to prevent the evaporation of large quantities of helium.

Several methods have now been devised for the protection of superconducting systems.

3.2.   Transformer Method

The transformer method is used both for superconducting systems working in the "frozen-current" mode, i.e., with the external supply source disconnected from the system, and for systems directly connected to the external supply. Protection is effected by means of a secondary circuit of normal metal inductively coupled to the superconducting coil.

It should be noted that in practical constructions the superconducting coil is almost always inductively coupled to various secondary circuits such as the metal walls of the cryostat, the metal framework of the winding, a shunting substrate of normal

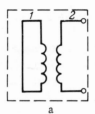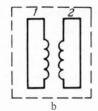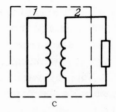

Fig. 3-3. Possible ways of connecting the protective winding. 1) Superconducting coil; 2) protective winding.

metal, and so on. The magnetic coupling coefficient between the
primary (superconducting) and secondary (normal) circuits is
given by the well-known equation

$$k = M_{12} / \sqrt{L_1 L_2}, \qquad (3\text{-}1)$$

where M_{12} is the mutual inductance between the primary and
secondary circuits, while L_1 and L_2 are the self-inductances of
these circuits. If the circuits are wound in the bifilar manner,
we may consider to a first approximation that $M_{12} = L_1 = L_2$ and
hence $k = 1$.

Experiments carried out with bifilar-wound superconducting
and normal solenoids [37-39] have revealed the principal qualita-

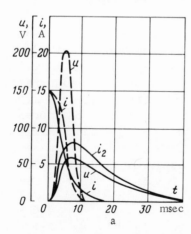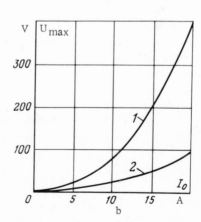

Fig. 3-4. Characteristics of the transition of the superconducting solenoid
into the normal state. 1) Protective winding open; 2) protective winding
closed.

tive laws of the transient processes for various modes of connecting the secondary (normal) circuit.

There are three possible ways of connecting the protective
winding: (a) winding open (Fig. 3-3a); (b) winding short-circuited
(Fig. 3-3b); (c) winding closed through an extra external resistance outside the cryostat (Fig. 3-3c).

Of course the arrangement of Fig. 3-3a corresponds to the
unprotected solenoid. Let us call the resistance of the protective
solenoid R_{ps} and the external resistance connected into its circuit
R_{ext}. In the arrangement of Fig. 3-3a $R_{ext} = \infty$, in Fig. 3-3b
$R_{ext} = 0$, and in Fig. 3-3c $0 < R_{ext} < \infty$.

Figure 3-4 shows the results of an experimental investigation
into a small superconducting solenoid [37], in which the coil was
wound at the same time as a copper wire coil (i.e., in the bifilar
manner). The superconducting coil was made of 65BT wire 0.25
mm in diameter and the protective coil of 0.1 mm copper wire; the
solenoid framework was dielectric; the internal diameter of the
winding was 16 mm and the external 34.5 mm, with a height of
35 mm; the number of turns in each coil was 2700, and the critical
current 20.1 A. In Fig. 3-4a the continuous curves relate to the
short-circuited protective coil and the broken curves to the unprotected superconducting solenoid (protective coil open) for an
initial solenoid current of 15 A. Figure 3-4b shows the maximum
potential difference U_{max} as a function of the initial current in the
superconducting solenoid I_0.

We see from these figures that the use of a short-circuited
protective coil leads to a sharp decrease in potential difference
over that part of the winding which has passed into the normal
state. The total duration of the transition process also increases
substantially. We see from Fig. 3-4b that the greater the initial
current in the solenoid (i.e., the higher the value of the stored energy) the more efficient the use of the protective coil becomes.

Figure 3-5a shows the time dependence of the instantaneous
power of the heat evolution in the solenoid for an open (broken
curve) and closed secondary circuit (w_c and w_M are the instantaneous power of heat evolution in the primary and secondary circuits). The same figure gives the integrated energy liberated in
the form of heat up to a particular instant in the primary (E_c) and
secondary (E) windings. Figure 3-5b respresents the maximum
instantaneous power of heat evolution in the solenoid as a function

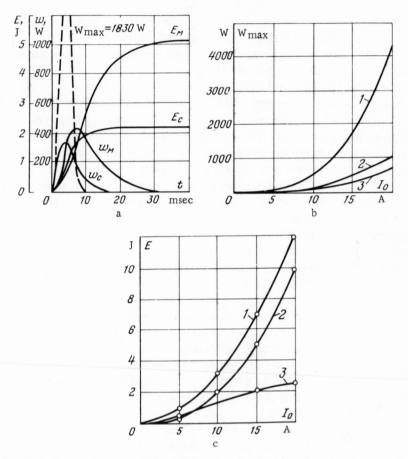

Fig. 3-5. Power of heat evolution and integrated energy as a function of time and
initial current.

of the initial current flowing in the latter for an open (curve 1) and
closed (curves 2 and 3) secondary circuit. Curve 2 relates to the
secondary circuit and curve 3 to the primary. Figure 3.5c shows
how the total energy stored in the solenoid and liberated in the
section of the superconductor coil passing into the normal state
depends on the initial current for an open protective circuit (curve
1). Curves 2 and 3 characterize the energy liberated in the secondary and primary circuits for a closed secondary circuit.

We see from the curves that the use of a short-circuited
secondary winding leads to a considerable reduction in the instan-

taneous power of heat evolution w_{max} in the section of the superconducting coil which has passed into the normal state. The
greater the initial current in the superconducting coil, the more
sharply w_{max} diminishes and the greater is the proportion of the
stored energy liberated in the protective winding. It should be
emphasized that, whereas in the primary (superconducting) circuit
the energy is liberated over a comparatively small section of the
winding (the part which has passed into the normal state), in the
secondary (protective) winding the energy is liberated uniformly
over the whole volume of the coil. The proportion of stored energy liberated in the secondary circuit is greater the smaller the
resistance of this coil. Clearly, when the resistance of the protective circuit is infinitely great, i.e., the secondary circuit is broken,
no energy is released in this circuit at all. A reduction in the resistance of the secondary circuit involves an increase in the cross
section of the secondary winding, and hence a reduction in the occupation factor of the winding relative to the superconducting material.

Thus the use of a short-circuited secondary coil inductively
coupled to the primary (superconducting) coil operating in the
"frozen current" mode is an effective means of reducing overvoltages and heat liberation in that part of the superconducting coil
which has passed into the normal state. However, all the stored
energy is liberated in the superconducting and protective coils inside the cryostat. This method does not therefore solve the problem of removing the energy from the cryostat.

In order to take a proportion of the stored energy outside the
cryostat we use an arrangement in which the protective (secondary)
winding is closed through a resistance R_{ext} situated outside the
cryostat. For $R_{ext} = \infty$ and $R_{ext} = 0$ all the stored energy is liberated inside the cryostat (in the first case in the normal part of
the superconducting coil, in the second case in the whole volume
of the protective winding as well as the normal part of the superconducting coil). Clearly there should be an optimum value $0 <
R_{ext} < \infty$ for which the maximum amount of energy is taken outside
the cryostat.

The greater the value of R_{ext}, i.e., the greater the total resistance of the protective circuit $R_{pc} = R_{ext} + R_{ps}$, the smaller is
the proportion of energy liberated in this circuit. In actual fact,
as already indicated, since the inductances of the primary and

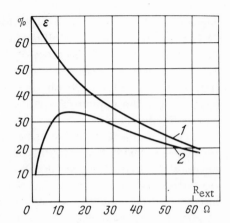

Fig. 3-6. Relative proportions of energy liberated in the secondary circuit
in relation to the active resistance lying
outside the cryostat.

secondary circuits are identical, the emf arising in the two circuits
during the transitional process are equal to one another, and hence
the heat-liberation power in the primary and secondary circuits
is distributed in inverse proportion to their resistances.

We see from Fig. 3-6 [38] how the relative proportions of the
total stored energy liberated in the secondary circuit as a whole
and in that part of the secondary circuit lying outside the cryostat,
i.e., the resistance R_{ext}, depend on the value of R_{ext} (curves 1
and 2, respectively). We see from these curves in fact that as
R_{ext} increases, the proportion of energy carried outside the cryostat passes through a maximum at a particular value of R_{ext}.

Figure 3-7a shows the time dependence of the potential difference at the ends of the section of superconducting coil which

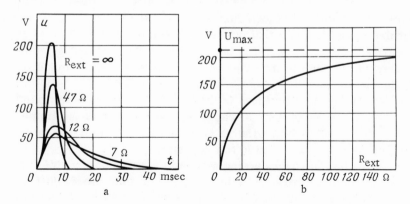

Fig. 3-7. Voltage in the superconducting coil as a function of time and R_{ext}.

has passed into the normal state for various values of R_{ext}. With increasing R_{ext} the overvoltage in the superconducting coil increases and the total duration of the transition shortens. The maximum voltage in the superconducting coil is shown as a function of R_{ext} in Fig. 3-7b (the broken line corresponds to the value of U_{max} for the unprotected solenoid, i.e., for a disconnected secondary circuit).

Thus, for a superconducting magnetic system working in the "frozen current" mode, taking part of the stored energy out of the cryostat (in order to reduce the evaporation of helium during the transition) involves worsening the conditions governing the behavior of the superconducting coil during the transition process. In fact, the overvoltage and the instantaneous power of heat liberation in the section of superconducting coil which has passed into the normal state then increase, and so does the proportion of the total energy liberated in this section. Thus, when the main protection problem lies in reducing the overvoltage and heat liberation in the section of superconducting coil in the normal state, protection may reasonably be provided in the form of a short-circuited secondary circuit. Our earlier discussion shows that the protection of the superconducting magnetic system using the transformer method is the more effective, the smaller the resistance of the secondary circuit R_{pc} and the closer to unity the magnetic coupling coefficient between the primary and secondary circuits. The secondary circuit may be made in the form of interlayers of normal metal (copper, aluminum) placed between layers of the superconducting coil. These interlayers form a short-circuited ring in which a current is induced during the transition process. It should nevertheless be noted that during the excitation of such a system a certain energy dissipation takes place in these windings or interlayers, leading to the evaporation of an additional amount of helium.

When the main problem is that of taking energy outside the cryostat (in the interests of reducing consumption of the liquid helium), the secondary winding is connected to a resistor or condenser (Section 3.3) outside the cryostat.

When the superconducting magnetic system is made of two or several sections working in the "frozen current" mode, each of these sections plays the part of secondary coil with respect to any other. When one of the sections passes into the normal state, its

current decreases, but at the same time the current in the other
sections inductively coupled to the first rises. Then the current
in these sections may exceed the critical value and a normal zone
will develop in these also. Early superconducting solenoids with
sectioned coils exploited this effect. The development of normal
zones in sections inductively coupled to that which first passed
into the normal state leads to a more uniform distribution of heat
evolution through the volume of the solenoid.

3.3. Discharge into an External Load

The protective method under consideration is used when the
superconducting magnetic system is connected to an external supply source. The simplest means of protecting such a system is
as follows. At the beginning of the transition into the normal state,
i.e., at the first signs of a normal zone appearing in the superconducting coil, the system is disconnected from the supply source
and discharged into a resistance R_{ext} lying outside the cryostat
(Fig. 3-8a). Clearly the greater the value of R_{ext} by comparison
with the resistance of the section of superconducting coil which
has passed into the normal state the greater the proportion of energy stored in the system which will be liberated outside the cryostat. An ideal means of protecting such a system is the interruption of the circuit ($R_{ext} = \infty$). However, R_{ext} is limited by the
fact that, in order to avoid the breakdown of the insulation (the
development of an arc between the current leads or between the
contacts of the disconnecting relay), the voltage in the current
leads at the instant of interruption (which is obviously equal to
$I_0 T_{ext}$, where I_0 is the initial current in the coil) should not be too
great.*

The greater the value of R_{ext}, i.e., the more energy taken
outside the cryostat, the smaller is the amount of heat liberated
in that section of the superconducting coil which has passed into
the normal state, and hence the slower the normal zone is propagated along the superconducting coil. In other words, the greater
the external resistance, the more slowly the resistance of the
superconducting coil R_{sp} increases. On the other hand, the larger

---

* At the present time in large superconducting magnetic systems protected in this way
the insulation of the current leads is usually designed for a voltage of up to 5000 V.

the value of R_{ext}, the more rapidly the current decreases in the circuit, since

$$I = I_0 e^{-t/\tau},\qquad(3-2)$$

where the time constant $\tau = L/(R_{sp} + R_{ext})$ decreases with increasing R_{ext}. Clearly therefore the greater the external resistance the smaller the potential difference will be in the section of the superconducting coil passing into the normal state ($U = IR_{sp}$).

It should be emphasized that the efficiency of this protection system increases with decreasing inductance of the superconducting magnetic system, since the time constant τ is directly proportional to the inductance. For a large inductance (and hence a large time constant) the energy will be taken out of the system too slowly, and hence a greater proportion of the energy will be liberated in the winding. To a first approximation the relationship between the amount of energy liberated in the coil (E_c) and that taken outside the cryostat (E_{ext}) is given by the following equation [40]:

$$\frac{E_c}{E_{ext}} \approx \left(\frac{\tau_l}{\tau}\right)^2,\qquad(3-3)$$

where $\tau_l = L/R_{ext}$ and is the time constant for the unprotected transition (when $R_{ext} = 0$). Clearly for a superconducting solenoid with a normal zone propagating inside it there is no point in determining τ in the form of $\tau = L/R_{sp}$, since R_{sp} increases with time. Hence τ should be defined as the time during which the initial current in the solenoid decreases by a factor of e.

On closing a small superconducting solenoid with an inductance $L = 1$ H and a time constant $\tau = 20$ msec into an external resistance

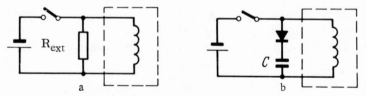

Fig. 3-8. Arrangements for discharging the superconducting system into an external load.

DISCHARGE INTO EXTERNAL LOAD

of 200 Ω, some 95% of the stored energy may be taken out of the cryostat.

Instead of an active resistance, a reasonably large-capacity condenser C may be used in this protection system (Fig. 3-8b). After the supply source has been disconnected and the superconducting solenoid connected to the condenser, current oscillations with a frequency of $\omega = 1/(LC)^{1/2}$ arise in the solenoid—condenser circuit. By using a diode in series with the condenser we may interrupt this oscillatory process in the first quarter of a period. As a result of this, the energy formerly stored in the superconducting magnetic system will be concentrated in the condenser.

From the point of view of the rate of taking energy out of the superconducting system, this protective arrangement becomes more effective the smaller the capacity of the condenser, since the smaller the value of C the greater is the frequency of the current oscillations in the circuit. On the other hand, however, a reduction in the capacity of the condenser leads to an increase in the voltage in that part of the winding which has been converted into the normal state. Actually the energy of the condenser is

$$E_{\text{cond}} = CU^2/2, \tag{3-4}$$

where U is the voltage on the condenser plates, equal to the voltage in the superconducting solenoid at the end of its transition into the normal state. Hence for the same amount of stored energy taken out of the solenoid and concentrated in the condenser, U will be the greater the smaller the value of C. Hence the minimum value of C will be determined by the permissible voltage in the current leads of the solenoid which has passed into the normal state.

On discharging into the active impedance (resistance) the voltage on the current leads of the solenoid is greatest at the beginning of the process, when the current in the system has its greatest value ($U_{\text{max}} = I_0 R_{\text{ext}}$). If the solenoid discharges into a condenser the voltage in the current leads assumes its greatest value on completion of the transient process, when the current in the circuit has fallen to zero and the accumulation of energy in the condenser has ended. From the point of view of possible breakdown in the insulation between the current leads of the superconducting system, protection by discharging into an active resistance is more dangerous, since breakdown of the insulation may then occur when the maximum current has developed in the system.

With regard to the effectiveness of the protection systems considered for reducing the overvoltage in the section of winding which has passed into the superconducting state, the use of an active impedance is preferable to that of a condenser. Actually, since at the beginning of the process there is no voltage on the condenser plates, in this period the energy is evacuated more slowly from the solenoid, and the current in the circuit decreases more slowly than in the case of an active resistance. The slow current decrease in the circuit leads to a faster extension of the normal zone along the winding due to the greater intensity of heat evolution, and hence to high potential differences in the normal part of the solenoid winding.

Clearly, from the point of view of the rate of evacuation of energy from the crystat, the most effective means of protection is that process in which the energy is led out at a constant voltage equal to the maximum permissible value U on the current leads (as determined by the electrical strength of the insulation). Since the current in the circuit decreases as the solenoid is being brought into the normal state, in order to satisfy the condition U = const we must arrange for a corresponding rise in the external active impedance during the transition. During the discharge of the solenoid after its disconnection from the supply source, the voltage may be kept approximately constant by using elements such as a nonlinear resistance, a previously charged condenser of large capacity, a storage battery withstanding large current pulses, and so on.

These considerations are illustrated in Fig. 3-9 by the time dependence of the current in the circuit of one of the experimental solenoids connected to an external supply source. The curve (based on experimental data [41]) shows that the most effective protection method is that based on a discharge at constant voltage.

One of the most promising ways of realizing this method of protection is the discharge of the superconducting magnetic system into an inverter-type converter which transfers the energy stored in the system to the ac network. Inversion of the stored energy may be effected either for a constant mean value of the power or for a constant mean value of the voltage at the ends of the solenoid equal to the mean value of the countervoltage (the counter-emf of the inverter).

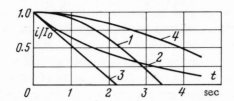

Fig. 3-9. Time dependence of the solenoid current for various discharge arrangements.   1)
Through a condenser; 2) through an active impedance (resistance); 3) through a unit maintaining constant voltage; 4) through the circuit without having the supply source disconnected.

It was mentioned earlier that if the voltage at the ends of the
solenoid is kept constant during the extraction of energy from the
solenoid the rate of energy extraction is greater than that achieved
in any other way. It is accordingly quite clear that the second
of the proposed modes of inversion is of special interest as a
means of protecting large superconducting magnetic systems designed for various purposes in the case of an accidental transition
into the normal state. In this case the inverter is used as an external load.

By comparison with the load resistances usually employed
as protective devices in such cases, semiconductor inverters are
incomparably more compact and require no powerful cooling system. These advantages are associated with the fact that only a
small proportion of the energy extracted from the magnetic system is dissipated in the inverters; up to 99% of the energy stored
in the solenoid is passed into the electrical network. This is very
important because, as already noted, in modern superconducting
magnetic systems the stored energy reaches the order of 10^8 J,
and systems of the order of 10^9 J are now being prepared.

In order to estimate the efficiency of the protection method
under consideration, we studied the discharge of a superconducting
solenoid into an inverter [42]. The results confirmed that the use
of an inverter was an extremely efficient means of evacuating the
energy stored in a superconducting magnetic system when the latter
accidentally passed into the normal state.

Let us consider one further combined system for the protection
of a superconducting magnetic system connected to a supply source.

In this system the coil of the superconducting system is inductively coupled to a short-circuited secondary winding of low resistance. The secondary winding is made in such a way that its inductance equals that of the primary (superconducting) coil and the magnetic coupling coefficient of the winding is nearly equal to unity.

When a normal section appears in the superconducting coil, the superconducting magnetic system is disconnected from the supply source. The current in the winding drops to zero almost instantaneously, while the energy dissipated in the system is absorbed by the secondary coil.* When the current ceases in the primary, a current is induced in the secondary. This induced current maintains the magnetic field. Thus the short-circuited secondary coil assumes the function of maintaining the magnetic field in the system. Since the resistance of this coil is nonzero, some of the stored energy is released in the secondary as Joule heat. However, since the resistance of the coil is low (and hence the time constant $\tau = L/R$ large) the current in the secondary winding and hence the magnetic field sustained by this winding decreases relatively slowly.

After a certain time, as a result of heat transfer to the liquid helium, the temperature of the superconducting coil decreases below the critical point for the current magnetic field, and this coil again passes into the superconducting state. After this the superconducting coil is again connected to the supply source and the current "transfers" from the secondary short-circuited coil to the primary. The supply source excites a current in the primary coil circuit, and the current in the secondary falls to zero.

Nearly all the protection systems discussed have one common failing. Even if the magnetic system (both winding and cryostat) remains intact, it ceases to fulfill its main function — generating a magnetic field. The magnetic field vanishes when the current in the superconducting coil diminishes. The only exception is that just considered, i.e., the system protecting the superconducting coil (connected to the supply source) by means of a short-circuited secondary winding inductively coupled to it. In this case, as already noted, the secondary winding assumes the function of maintaining the magnetic field. However, because of the necessity of ensuring a low resistance of the secondary winding, the scale di-

---

* As a result of this no overvoltages arise in the current leads of the superconducing coil.

mensions of the latter become comparable with those of the superconducting system itself. This method of protection cannot therefore be recommended for extensive use.

However, in a number of technical systems using superconductors to produce magnetic fields, the vanishing of the field may itself lead to an emergency, even if the magnetic system itself remains undamaged. Clearly the foregoing methods of protecting
magnetic systems based on superconductors are far from perfect.

We indicated earlier that the accidentally induced transition
of the superconducting magnetic system into the normal state was
the result of the formation of a normal zone in the winding. The
the ideal method of protection would be one in which the possibility
of a normal zone developing would be excluded, or at least its uncontrolled propagation along the coil would be prevented if such
a normal zone did in fact appear. The stabilization of superconducting coils which we shall subsequently consider is a method of this
kind.

3.4. Reasons for the Development
of a Normal Zone

Before approaching the question of stabilizing superconducting
magnetic systems, let us give a little more detailed consideration
to the reasons underlying the creation of a normal zone in the
winding.

It was indicated in Chapter 2 that the properties of superconducting materials may be characterized by three main parameters:
the critical temperature $T_c(B, I)$, the critical magnetic induction
$B_c(T, I)$, and the critical current $I_c(B, T)$. In other words, the
superconducting region is characterized by a certain function
$G_c(T, B, I)$ characteristic of the particular material. The reliable
operation of a superconducting device is only possible if
$G(T, B, I) < G_c$. On exceeding G_c the superconducting properties
of the material vanish and a normal zone appears in the winding.

Clearly detailed information regarding the properties of the
superconducting material used in the winding would enable us (in
principle) to choose working conditions of the magnet that would
prevent the development of parameters exceeding the critical
values at any point of the winding, and so eliminate all possibility
of the appearance of a normal zone. However, in practice the

position is complicated by the fact that, by virtue of various defects in the technological process, the superconducting wire does
not always have the same properties along its entire length. The
greater the length, the greater is the probability that a particular
section of wire will contain a "weak" region in which the critical
properties of the material are lower than those of the remainder.
The critical properties of the coil made from this piece of wire
will as a whole be determined by the properties of the "weak"
part. With respect to the determination of the critical parameters
of the wire in a short sample, the probability that a "weak" part
will occur in this particular sample is quite low. The larger the
magnet, i.e., the more wire required for constructing the coil, the
greater is the probability that "weak" sections will appear in the
winding.

We should also mention another cause of the development of
a normal zone in the coil, which is similar in nature to that already
indicated. In view of the fact that it is practically impossible to
fabricate the coil of a large superconducting device from a single
piece of wire (this being determined by the maximum length of the
wires currently manufactured) the problem of connecting individual
wires together becomes very important. However, for a large number of joints it is unreasonable to expect a perfect superconducting connection. In a coil made up of several series-connected
wires, Joule heat is therefore usually evolved at the joints, and
this may lead to considerable local heating of the superconductor,
raising its temperature above the critical and creating a normal
zone.

Situations promoting the development of a normal zone in the
coil may arise during the practical use of superconducting magnets.
A particular case in point is the displacement of the turns in the
coil of the magnetic system. On increasing the current, in fact,
the forces acting on the coil carrying the current in a magnetic
field are augmented. If the turns of the coil are poorly secured,
at a certain current they will begin moving. When the current-
carrying conductor moves in the magnetic field, a certain amount
of energy will be dissipated. The part of the coil containing the
turn in question will be heated to a temperature exceeding the critical, and a normal zone will appear in the superconducting coil.

As indicated in Section 2.2, the magnetization of nonideal
superconductors of the second kind is essentially an irreversible

process; a certain amount of energy is evolved during their magnetization. This heat evolution may lead to the heating of the coil
above the critical temperature and hence to a transition into the
normal state.

The least predictable circumstance leading to the appearance
of a normal zone in the winding, and therefore the most dangerous,
is the abrupt penetration of a magnetic field into a nonideal superconductor of the second kind (flux jump).

As indicated in Section 2.3, under certain conditions (relatively large wire diameter, low temperatures, region of weak magnetic
fields) a flux jump may become "catastrophic," i.e., lead to the
appearance of a normal zone in the superconducting wire. The
less favorable the conditions of heat transfer from the portion of
the coil under consideration to the helium coolant, the more likely
is the appearance of a "catastrophic" flux jump. The conditions
of heat transfer from a turn of wire situated in the inner layers
of the coil are much worse than those corresponding to an individual sample of wire immersed in a helium bath. Thus, if we consider a short wire sample and a coil carrying the same current
and assume that a flux jump of equal intensity takes place in the
short sample and in the coil, it may well be that the jump will not
be "catastrophic" for the short sample, but will be so in the coil
because of the interior conditions of heat transfer.

One of the most serious problems encountered some years ago
by the fabricator of the first superconducting magnetic systems was
the "degradation" of the current in the superconducting coils. This
is due to the fact that the critical currents found acceptable in a
superconducting coil* were much weaker than the critical currents
characterizing the same superconducting wire, measured by tests
on short samples.

In the first attempts at the creation of superconducting magnetic systems, when it had been found that the acceptable current
density in any winding diminished in proportion to its size, this
factor was the cause of great pessimism regarding the prospects
of making any large superconducting systems at all. It was found
that with increasing size of the coils their critical currents decreased to negligibly small values. This is illustrated in Fig. 3-10,

------

*By critical current of the coil we mean the current which causes the coil to revert to
the normal state.

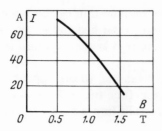

Fig. 3-10. Critical current as a function of the magnetic induction of a solenoid.

which shows the change in critical current for a series of geometrically similar solenoids (Z. Stekley, 1963 [42]). These solenoids were fabricated from niobium – zirconium wire and had internal diameter, external diameter, and height in the ratio of 1 : 1.8 : 0.6; the density factor of the winding relative to the superconducting material was 0.3. The internal diameter of the solenoids in this series varied from 10.5 to 153 mm. In Fig. 3-1 0 the horizontal axis shows the induction of the magnetic field obtained in the solenoids; of course, in geometrically similar solenoids the induction is proportional to the solenoid dimensions.

Subsequently developed methods of combating current degradation by stabilizing the superconducting coils (these being the subject of our present book) made it possible to overcome these difficulties.

Starting from 1962, various research workers one after the other proposed different explanations for current degradation in superconducting systems. The first attempts were based on a purely probability approach, based on the existence of a "weak" section of superconducting wire such as we have just considered. However, it was subsequently shown that the explanation of current degradation in this way involved irresolvable contradictions. An attempt at explaining the effect by the displacement of the turns of the coil was also incapable of establishing the reasons for all the phenomena associated with current degradation.

An explanation for this effect based on the possible transition of the winding into the normal state due to heat evolution during magnetization was also unable to account for these degradation effects in their entirety. However, this explanation was closer to the presently accepted treatment of the effect than all the others. The explanation now most widely accepted for degradation is based on the flux-jump mechanisms as causing the development of a normal zone in the coil. This point of view regarding the nature of the current degradation effect is evidently the nearest to the true state of affairs. In any case methods of combating degradation based on this concept (methods of stabilizing superconductors) have by far led to the best results.

3.5. Stabilization of Superconductors
in Their Various Forms

It was mentioned in Section 3.3 that the best method of protecting superconducting magnetic systems was stabilization. Stabilization of superconductors means a system of protective measures aimed at reducing the probability of the development of a normal zone, or preventing the extension of this zone along the winding if for some reason it should arise.

Known methods of superconductor stabilization may be distinguished according to the problems which they solve; the two main groups are thermal (or cryostatic) and internal stabilization.

The thermal stabilization of a superconductor eliminates the propagation of the normal zone along the coil when a normal nucleus appears within the latter, whatever the reasons for its appearance (a "catastrophic" flux jump, the motion of the turns of the coil under the action of magnetic forces, the overheating of contact resistances, etc.).

Internal stabilization solves only part of this problem; it prevents the passage of the superconducting coil into the normal state due to a flux jump $I < I_c$, and hence ensures that the same current density will be attained in the winding of the system as in a short sample. However, if a normal zone appears in the winding not as a result of a flux jump, but for some other reason, internal stabilization cannot prevent the inevitable transition of the whole winding into the normal state. Thus, the method of internal stabilization is less universal than the method of thermal stabilization.

Different methods of stabilization are realized by using different constructions of the conductors for the winding of superconducting devices. We shall subsequently give the name "combined conductor" to one in which some of the cross section is occupied by superconducting material (a nonideal superconductor of the second kind) and the rest by a normal metal (copper, aluminum, etc.). A combined conductor is ordinarily used either in the form of a cable in which the superconducting filaments are twisted together with filaments of a normal metal and made monolithic with a filler of normal metal (such as indium), or else in the form of a matrix of normal metal with one or several superconducting strands inside it.

The method of thermal stabilization lies in shunting the superconductor with a certain amount of normal metal having good thermal and electrical conductivities. If a normal zone is created in the coil, the current passing through the superconductor is transferred into the shunting metal and passes around the normal section. For the particular conditions which we shall be considering, the normal zone vanishes ("collapses") after the reasons for its appearance have been eliminated. The superconductor returns to the superconducting state and is again capable of supporting the transport current.

The methods of internal stabilization are directed at preventing the temperature of the superconductor from rising to $T_c(B, I)$ as a result of heat evolution due to flux jumps. This is achieved either by reducing the value of this heat evolution or by increasing the heat capacity of the conductor relative to the heat evolved.

The first method (we shall call this intrinsically internal stabilization) lies in breaking the superconductor down into very fine filaments [43]. Since the intensity of the flux jump diminishes with declining diameter of the superconducting filament (Chapter 2), for a certain filament diameter (usually a few microns) a "catastrophic" flux jump becomes impossible, or at least the energy released in this fashion is insufficient to raise the temperature of the superconductor to the critical.

Another method of internal stabilization – enthalpy (or adiabatic) stabilization – is realized by including components of great volumetric heat capacity (such as lead) into the conductor [44]. In this case the conditions of heat storage are improved and the maximum temperature reached as a result of heat evolution during the flux jump is reduced.

There is yet another way of limiting the maximum temperature of the conductor during heat evolution arising from a flux jump. If the rate of this heat evolution is reduced (the evolution of the same amount of heat extended in time), then for the same rate of heat outflow the maximum temperature of the conductor will be less than in the case of the ordinary rate of heat evolution.

The idea of "extending" the flux jump forms the basis for the "electrodynamic stabilization" method of Chester [45]. During the redistribution of the magnetic induction over the cross section of the superconductor which occurs after the flux jump, in accordance with the Le Chatelier—Braun principle [36] factors opposing this redistribution arise, that is, eddy currents (damping the induction redistribution process) are induced in the normal metal of the combined conductor. Thus the higher the electrical conductivity of the normal component of the combined conductor the more slowly do these eddy currents die out, and the more flux jump is "extended" in time. Clearly in order to realize this stabilization method a normal metal with the minimum resistivity (copper or very pure aluminum) should be introduced into the conductor.

The method of transient stabilization may also to some extent be regarded as an internal-stabilization method. This is to a certain degree similar to the method of enthalpy stabilization, but, whereas in the latter case the heat-retaining components are introduced into the composition of the combined conductor, in transient stabilization the heat-retaining material is not a constituent part of the actual conductor. Interlayers of a material having a fairly large specific heat (copper, aluminum, lead, cadmium) are placed between layers in the winding of the superconducting magnetic system [46, 47]. The main idea of this stabilization method lies in the fact that the thermal flux from the conductor to the heat-retaining interlayer may exceed the critical thermal flux to the helium tank. Such an intense heat outflow reduces the maximum temperature reached as a result of the heat evolution accompanying the flux jump. The transient stabilization method has a number of common features with thermal stabilization.

It should be emphasized that the various methods of stabilization here discussed, taken in isolation, are to a certain extent abstract, since in actual fact aspects constituting the typical features of each stabilization method often overlap. In practice, furthermore, several such methods are frequently used together, rather than one at a time.

The use of these various methods of stabilizing superconducting coils does not supplant the use of the protection systems considered earlier. In using large superconducting magnetic systems with a great deal of stored energy there may well be emergency situations (for example, a loss of cooling by a part of the winding) requiring the rapid evacuation of energy from the magnetic system. In actual fabrication a complex of protective measures is therefore employed, including various methods of stabilization and also methods preventing the breakdown of the coil in the case of its unforeseen transition into the normal state.

Part II

METHOD OF THERMAL STABILIZATION

Chapter 4

Equilibrium of the Normal Zone in Combined Conductors Under Isothermal Conditions

4.1. Stekly Model of a Stabilized Superconductor

Thermal (cryostatic) stabilization is the name given to a system of protective measures preventing the transition of a superconducting device into the normal state when a normal zone appears in the superconducting coil.

As already indicated, the basic idea of the method of thermal stabilization lies in reinforcing the superconductor with a normal metal possessing high electrical and thermal conductivities. The superconductor is thus shunted by the normal metal, and if a normal zone appears the current flowing through the superconductor "bypasses" this zone through the shunt (usually called the substrate).

The principle of the thermal stabilization of superconductors was proposed by Stekly in 1965 [48]. It should be emphasized that even before the appearance of this proposal general considerations made it obviously desirable to use a superconductor reinforced with copper; the production of superconducting conductors with a copper coating was started even in 1962. However, only when the Stekly paper appeared were the foundations of the thermal-stabilization method first clearly defined, and this became the origin of a new and fruitful direction in the creation of large superconducting magnetic systems.

As already mentioned, the appearance of a normal zone in a superconducting coil may arise from various causes. The most unpredictable and hence most dangerous cause from the point of view of the service conditions of the magnetic system is a "catastrophic" flux jump.

If a normal zone is formed in a superconducting conductor not having a normal metal substrate, Joule heat will be evolved as the current flows through the part converted into the normal state. This heat will be carried away by heat transfer to the ambient (liquid helium bath) and by conduction to the neighboring superconducting parts of the conductor. Under the influence of the heat flow the temperature of the neighboring parts will rise, and on reaching the temperature $T_c(B, J)$ they also will pass into the normal state. There will thus be a continuous propagation of the normal section, which will continue until the entire winding has reverted to the normal state.

When the current flows through a combined conductor (a superconductor in a matrix of normal metal), the physical processes taking place during the development of the normal zone in the superconductor will be very different.

The analysis which we shall now present for these processes will be carried out for the case in which the reason for the development of the normal zone is an increase in the current passing through the conductor so as to rise above the critical value of $I_c(B, T)$. As we shall shortly see, the results of the analysis will also extend to cases in which the appearance of the normal zone arises from any other cause: a "catastrophic" flux jump, movement of the turns in the winding, local overheating, and so on.

Let us consider a combined conductor consisting of a superconductor and a shunt (substrate) composed of normal metal (Fig. 4-1). We shall consider that the conductor exists in a steady magnetic field with induction B and is placed in a liquid helium tank having a constant temperature T_t. Let us make the following assumptions:

1. The temperature is constant over the cross section of the combined conductor and equal to the temperature at its surface;

2. Ideal thermal and electric contact exists between the superconductor and the substrate;

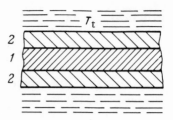

Fig. 4-1. Structure of a combined
conductor. 1) Superconductor; 2)
substrate.

3. The electrical resistance of the superconductor in the normal state is very large compared with that of the substrate
   material, and hence if the superconductor is in the normal
   state all the transport current will flow through the substrate.*
4. The resistivity of the substrate material is independent of
   temperature;
5. The heat-transfer coefficient from the surface of the combined conductor to the liquid helium is independent of temperature;
6. The temperature of the part of the conductor under consideration is identical along its whole length, i.e., the transition of the superconductor into the normal state takes place
   at the same time all along the part under consideration.

   Let us denote the current steadily flowing through the combined conductor as I. Clearly while the superconductor remains
in the superconducting state the entire current will flow through
the superconductor and none through the substrate. In this case
there is no heat evolution in the combined conductor and its equilibrium temperature equals that of the helium tank T_t.

   Let us designate the critical current of the superconductor at
this temperature as I_{ct}; the critical current of this superconductor
at an arbitrary temperature T we will label as $I_c(T, B)$.

---

* At helium temperatures the resistivity of a standard electrotechnical copper of type
Cu 01 is 8×10^{-9} Ω-cm, while that of a Nb–60% Ti alloy wire at the same temperature is 220×10^{-6} Ω-cm. However, since the magnetoresistive effect increases for
copper in strong magnetic fields, the difference between the resistivities will be
slightly less. For Soviet copper type MM (All-Union State Standard 2112-62), usually
employed as a substrate material, $\rho_{4.2 \ K} \approx 15$-$20 \times 10^{-9}$ Ω-cm; thus in strong magnetic fields the resistivities of the substrate and the superconductor in the normal state
will differ by about four orders of magnitude.

It is obvious that

$$I_{ct} = I_c(T_t, B).\tag{4-1}$$

The critical temperature of the superconductor in the specified field at zero current is

$$T_{c0} = T_c(0, B).\tag{4-2}$$

It is well known that for the majority of nonideal superconductors of the second kind the temperature dependence of the critical current at B = const may be regarded as linear to an acceptable degree of accuracy. We may therefore write

$$\frac{I_c(T, B)}{I_{ct}} = 1 - \frac{T - T_t}{T_{c0} - T_t}.\tag{4-3}$$

We see from the foregoing that, if a current $I < I_{ct}$ flows through the combined conductor, the superconductor will remain in the superconducting state, all the current will flow through it, and the temperature of the combined conductor will equal the ambient (helium tank) temperature T_t. If the current in the combined conductor increases, then on reaching a certain value I_c the section of superconductor under consideration will transform into the normal state and its temperature will greatly exceed T_t. In accordance with the foregoing discussion the current I will be completely forced into the substrate, and none will then flow through the superconductor.

Since the substrate composed of a normal metal possesses a certain electrical resistance (although several orders of magnitude smaller than that of the superconductor in the normal state), when the current I_c flows through the substrate every unit length of the conductor will liberate Joule heat

$$Q_J = \rho I_c^2 / A,\tag{4-4}$$

where ρ is the resistivity of the substrate material and A is the cross section of the substrate.

Clearly this heat will pass to the helium tank, and thermal equilibrium will be established in the system at a temperature T, at which the thermal flux from the surface of the combined conductor into the helium tank becomes equal to the Joule heat evolution in the substrate:

$$hP(T-T_t) = Q_J,$$ (4-5)

where h is the heat-transfer coefficient from the surface of the conductor into the helium tank and P is the perimeter of the combined conductor.

The temperature at which this thermal equilibrium is established depends on the resistivity of the substrate ρ, the heat-transfer coefficient h, the substrate perimeter P, and the substrate cross section A. If ρ = const and h = const, T may be varied by varying A and P. Of course, the larger the proportion of normal metal in the combined conductor, i.e., the larger the value of A, the smaller is the Joule heat evolution. The greater the perimeter of the conductor P, the greater is the heat flow from its surface; hence the greater the values of A and P, the lower the equilibrium temperature T.

Depending on the values of A and P for specified ρ and h the following situations may arise.

1. If the amount of normal metal in the combined conductor is small, then after a current I_c has been reached the current will be forced into the substrate and thermal equilibrium will be established at a temperature of the combined conductor $T > T_{c0}$. In this case, the current will flow solely through the substrate. Allowing for (4-4) and (4-5), the thermal-balance equation will be written in the following form:

$$hP(T - T_t) = \rho I_c^2 / A.$$ (4-6)

Although the superconductor passes into the normal state at a current $I = I_c$, a subsequent reduction in current (i.e., the removal of the cause which led to the development of the normal state) will not restore the material to its superconducting state. This is because, over a certain range of currents $I < I_{ct}$, the Joule heat evolution in the substrate is so great that the equilibrium temperature T of the combined conductor will be greater than T_c. It is plain from Eq. (4-6) that this equilibrium temperature will become equal to T_c on reducing the current to the value

$$I = \sqrt{\frac{AhP(T_{c0} - T_t)}{\rho}}.$$ (4-7)

2. If the amount of normal metal in the combined conductor is increased and its surface is developed, then for certain values of A and P thermal equilibrium will be established at a combined-conductor temperature of $T = T_{c\,0}$. In this case, as in case 1, the superconductor is in the normal state and the total current I_{ct} flows through the substrate. The heat-balance equation has the form:

$$hP\,(T_{co} - T_t) = \rho I_{ct}^2/A. \qquad (4\text{-}8)$$

3. If we increase the values of A and P still further, thermal equilibrium will set in for a combined-conductor temperature of $T < T_{c\,0}$. This case requires detailed consideration. It was assumed earlier that after the current in the combined conductor had exceeded I_{ct} the superconductor passed into the normal state and all the current passed into the substrate. However, since in the case under consideration the equilibrium temperature of the combined conductor is lower than the critical temperature of the superconductor T_{c0} in the field specified at zero current, the superconductor will turn to the superconducting state. The only difference is that, whereas before the transition its temperature was equal to the temperature of the helium tank T_t, this temperature now lies in the range $T_t < T < T_c$.

Clearly if the superconductor has returned to the superconducting state it will again be able to carry a transport current; hence some of the current in the substrate will return to the superconductor. Since the proportion of current flowing through the substrate will then diminish, and hence so will the intensity of Joule heat evolution in the conductor, this return of current to the superconductor will continue until a new equilibrium temperature of the combined conductor has been established. Here we are only considering part of the current I_{ct}, since the entire current cannot return from the substrate; this would again return the superconductor to the normal state. Let us call the proportion of current remaining in the substrate f.

As indicated in Ref. 48, a proportion of the current I_c (equal to the critical current of the superconductor at the equilibrium temperature T of the combined conductor) will return to the superconductor

$$(1-f)I_{ct} = I_c(T,\ B), \qquad (4\text{-}9)$$

where $(1 - f)$ is the proportion of current returning to the super-conductor.

It would appear that the superconductor should thereupon again pass into the normal state, since the critical current will be flowing through it. However, in the present case the heat release from the combined conductor into the helium tank is so intense that the heat evolution in the superconductor due to the steady displacement of the vortices for a current of $I = I_c(T, B)$ in the superconductor will not lead to any rise in temperature. Hence in this case the superconductor is in a resistive state. As indicated in Section 2.4, in this state the superconductor possesses a certain resistance, and Joule heat is liberated when a transport current flows through it.

This explanation for the behavior of a combined conductor is extremely simple and schematic, but it enables us to represent the complex processes taking place inside the conductor when the latter passes a current in an explicit manner.

Let us consider the principal relationships for the case in which the current is divided between the superconductor and the substrate in a combined conductor. It follows from the foregoing considerations that, in general, the division of current between the superconductor and the substrate may take place not only for a current of I_c in the combined conductor but for an even weaker current (case 1). Let us assume that a current $I \leq I_{ct}$ flows along the combined conductor. This current is divided between the substrate and the superconductor; the current through the substrate is

$$I_{sub} = fI, \qquad (4\text{-}10)$$

that through the superconductor is

$$I_s = (1-f)I. \qquad (4\text{-}11)$$

Since the resistance of the substrate per unit length is equal to ρ/A, the decrease in voltage per unit length of the combined conductor is

$$U = fI\rho/A, \qquad (4\text{-}12)$$

while the Joule heat evolution per unit length of the combined conductor is

$$Q_J = UI = fI^2\rho/A. \tag{4-13}$$

This equation gives the Joule heat not simply in the substrate but in the total conductor, i.e., in the substrate and in the superconductor existing in the resistive state.

The heat-balance equation for this case may be written

$$hP(T-T_t) = fI^2\rho/A. \tag{4-14}$$

From Eq. (4-9) we have

$$(1-f)I = I_c(T, B). \tag{4-15}$$

Thus on allowing for (4-3) we have

$$(1-f)\frac{I}{I_c t} = 1 - \frac{T-T_t}{T_{c0}-T_t}. \tag{4-16}$$

After making some simple algebraical transformations Eqs. (4-12), (4-14), and (4-16) may be reduced to dimensionless form

$$f = \frac{\iota - 1}{\iota(1-\alpha\iota)}; \tag{4-17}$$

$$\upsilon = \frac{UA}{\rho I_c t} = \frac{\iota-1}{1-\alpha\iota}; \tag{4-18}$$

$$\tau = \frac{\alpha\iota(\iota-1)}{1-\alpha\iota}, \tag{4-19}$$

where

$$\iota = I/I_c t \tag{4-20}$$

is the dimensionless current in the combined conductor;

$$\tau = \frac{T-T_t}{T_c-T_t} \tag{4-21}$$

is the dimensionless temperature of the combined conductor; the quantity v (4-18) is the dimensionless voltage drop per unit length of conductor;

$$\alpha = \frac{\rho I_c^2}{AhP\,(T_c - T_t)} \qquad (4\text{-}22)$$

is a dimensionless parameter usually called the Stekly stability criterion.

The parameter α characterizes the properties of the combined superconductor, its construction, and the conditions of heat transfer to the helium tank.

If in the combined conductor only the amount of normal metal in the substrate, i.e., the values of A and P, are varied, the quantities I_c, T_c, ρ, and h naturally remain constant. The more normal metal the combined conductor contains, the smaller is the parameter α.

Equations (4-17) to (4-19) enable us to make a detailed analysis of the processes taking place in the combined conductor for various values of ι and α.

It should be emphasized that these equations were obtained for conditions under which the current is divided between the superconductor and the substrate. Only those values of ι for which this division occurs may therefore be substituted into the equations. By themselves the equations cannot determine the range of ι values within which current division occurs; as we shall later show, these values of ι have to be determined from other considerations.

Let us consider the relationship between the dimensionless voltage drop per unit length of the combined conductor $v = UA/\rho I_{ct}$ and the dimensionless total current in the combined conductor ι for various values of the parameter α, i.e., the volt−ampere characteristic of the combined conductor. Such a characteristic is shown for $\alpha = 1$ in Fig. 4-2.

It follows from Eqs. (4-17) and (4-18) that

$$v = f\iota. \qquad (4\text{-}23)$$

The case in which $f = 1$, i.e., in which all the current flows through the substrate, is represented by a straight line passing through the origin at an angle of $\pi/4$ in the system of coordinates

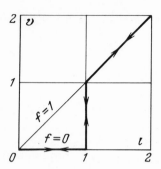

Fig. 4-2. Volt—ampere characteristics
of a combined conductor for $\alpha = 1$.

under consideration. This straight line represents a relationship
corresponding to Ohm's law:

$$U = IR,\qquad(4\text{-}24)$$

where $R = \rho/A$.

The case $f = 0$ in which all the current flows through the superconductor is reflected by points on the horizontal axis of the foregoing diagram, since no voltage drop occurs when current flows
through a superconductor in the superconducting state.

Let us first consider the case in which $\alpha = 1$. Clearly, if in
the combined conductor passing a current $I = I_{ct}$ thermal equilibrium is established at a temperature $T = T_{c0}$, Eq. (4-8) in dimensionless form will appear as $\alpha = 1$. Thus the combined conductor
characterized by a value of $\alpha = 1$ is one in which on passing a current I_{ct} (the critical current of the superconductor at the temperature of the helium tank T_t) thermal equilibrium is established for
a temperature of the conductor equal to the critical temperature
of the superconductor in question at zero current.*

Thus in this case for $I = I_{ct}$ (i.e., $\iota = 1$) all the current flows
through the substrate. For $I < I_{ct}$ the temperature of the combined
conductor remains below T_{c0}, the superconductor passes into the
superconducting state, and the total current $I < I_{ct}$ returns to the
superconductor.

---

*More strictly speaking, thermal equilibrium of the combined conductors may occur
at $\alpha = 1$ for the critical current I_c over the whole temperature range T_t to T_c.

In Fig. 4-2 the values of the dimensionless voltage decrease
for the case $\alpha = 1$ and $\iota \geq 1$ lie on the straight line $f = 1$ and for
$\iota \leq 1$ on the horizontal axis ($f = 0$); for $\iota = 1$ the quantity $UA/\rho I_{ct}$
changes from 0 to 1.

It follows from this that, if the reason for the appearance of
a normal zone in the superconductor is of a fluctuational character
(for example, a "catastrophic" flux jump), then in the combined
conductor in which $\alpha = 1$ superconductivity will be restored (for
any currents up to $I = I_{ct}$) immediately after the removal of the
cause which gave rise to the normal zone. In other words, in this
combined conductor the superconductor will remain in the superconducting state for any currents $I < I_{ct}$. Hence in such a combined conductor the effect of current degradation will be entirely
excluded. This kind of conductor is usually called completely
stabilized.

It is important to have a clear idea of the physical meaning
of the parameter α.

As already indicated, when a normal zone appears in the superconductor, at the initial instant all the current is expelled into the
substrate, and the subsequent "fate" of this current depends on the
conditions governing the release of the Joule heat evolved in the
substrate in these initial instants after the creation of the normal
zone. Thus it is important to evaluate the temperature of the combined conductor when the entire current I_c flows through the substrate. For this case the thermal balance equation may be written
as follows:

$$hP(T - T_t) = \rho I_{ct}^2/A, \qquad (4\text{-}25)$$

whence

$$\frac{T - T_t}{T_{c0} - T_t} = \frac{\rho I_{ct}^2}{AhP\,(T_{c0} - T_t)}, \qquad (4\text{-}26)$$

i.e.,

$$\tau = \alpha. \qquad (4\text{-}27)$$

It should be emphasized that this equation is only valid when
all the current flows through the substrate, this current being
critical (I_{ct}). Clearly if $\alpha > 1$ for $I = I_{ct}$ we obtain $T > T_{c0}$ and

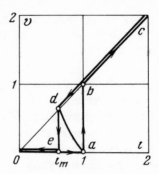

Fig.4.3. Volt—ampere characteristics
of a combined conductor for $\alpha > 1$.

all the current will remain in the substrate; only on reducing the
current below the value given by Eq. (4-7) will the temperature of
the combined conductor decrease below $T_{c\,0}$ and the current return to the superconductor. However, if $\alpha < 1$ then for $I = I_{ct}$ we
obtain $T < T_{c0}$ and the current will be divided between the superconductor and the substrate.

Let us now consider the case $\alpha > 1$ (Fig. 4-3). On increasing
the current I the superconductor will remain in the superconducting state until the current reaches the critical value $I = I_{c0}$. As
the current is increased while $I < I_{c0}$ the temperature of the combined conductor equals the temperature of the helium tank. On
exceeding the critical current the superconductor will pass into
the normal state and the whole current will be expelled into the
substrate. The temperature of the combined current will then be
above the critical temperature for zero current T_{c0}; this is obvious from Eq. (4-27). The states corresponding to current values
$\iota \geq 1$ lie on the straight line $f = 1$ in the volt—ampere characteristic. Thus, on increasing the current the combined conductor
with $\alpha > 1$ will pass through the states lying on the $O-a-b-c$
section (Fig. 4-3).

Let us consider the processes associated with a reduction in
the current starting from a certain value $\iota > 1$. For $\iota > 1$ all the
current flows in the substrate, in which a certain amount of Joule
heat is liberated. Let us reduce the current to $\iota = 1$ ($I = I_{ct}$). It
would appear that the superconductor should then return to the
superconducting state. However, this does not happen; for $\alpha > 1$

the current passing through the substrate even for $\iota < 1$ heats the conductor so much that its temperature rises above $T_{c\,0}$ for a certain range of currents $\iota < 1$. Of course the lower the current I ($I < I_{ct}$) the lower will the equilibrium temperature of the combined conductor T be. Finally, for a certain current $I_m < I_{ct}$ the equilibrium temperature T will become equal to $T_{c\,0}$; on further reducing the current the superconductor will pass into the superconducting state and the total current will return to the superconductor. Thus on reducing the current the combined conductor in which $\alpha > 1$ will pass through the states lying on the section $c-d-e-0$.

The quantity I_m is called the minimum current for the existence of the normal zone and is defined by Eq. (4-7), which on allowing for (4-22) may be written in the form

$$\iota_m = 1/\sqrt{\alpha}, \qquad (4\text{-}28)$$

where $\iota_m = I_m/I_{ct}$.

It follows from the foregoing that if the reason for the appearance of the normal zone in the superconductor has a fluctuational nature then in the combined conductor for which $\alpha > 1$ superconductivity will only be restored after removing the cause of the normal zone if $I < I_m$. In other words, in this combined conductor the superconductor will remain in the superconducting state for any currents $I < I_m$. However, in the current range $I_m = I_{ct}/\sqrt{\alpha}$ to I_c the superconductor which has passed into the normal state for reasons of a fluctuational character will not return to the superconducting state after the reason for the development of the normal zone has been removed. This is because, owing to Joule heat evolution in the substrate, the temperature of the superconductor will remain above T_c. Hence such a conductor is only stabilized for currents $I < I_m$; for larger currents it is nonstabilized.

Conductors only stabilized up to certain currents $I_m < I_{ct}$ are usually called partially stabilized.

Clearly the points a and d (Fig. 4-3) correspond for a particular value of α to the two possible limiting states of the combined conductor for which the current is divided between the superconductor and the substrate, i.e., states described by Eqs. (4-17) to (4-19). It follows from these equations, in fact, that one of the limits $f = 0$ for which the whole current flows through the superconductor occurs when $\iota = 1$ and in this case $v = 0$ (point a). The other limit $f = 1$, for which all the current flows through the substrate, occurs at a current

$$\iota_m = 1/\sqrt{\alpha}.$$

This is obvious from Eq. (4-17). The resultant value of ι_m naturally coincides with the value given by Eq. (4-28) for the point d.

Relative to the volt−ampere characteristics the line limited by the points a and d is the geometrical locus of the equilibrium states of the conductor corresponding to the division of the current between the superconductor and the substrate for $0 < f < 1$. This line, defined by Eq. (4-18), is shown in Fig. 4-3; along the line, $T_t < T < T_c$.

We see from Fig. 4-3 that along the line $a−d$

$$dU/dI < 0, \tag{4-29}$$

i.e., the total resistance of the combined conductor $R = dU/dI$ is negative in this range of conditions.

Let us give some more detailed consideration to the question of the possible realization of states corresponding to the section $a−d$. On increasing the current I in the combined conductor above I_{ct} the superconductor passes into the normal state and the current flows solely through the substrate. However, it is only possible to pass through the value I if the current source is able to provide a voltage per unit length of the combined conductor greater than $v = 1$. If after reaching the current I_c the voltage on the combined conductor is raised gradually from zero, the current in the conductor will decrease; there will then be a redistribution of the current between the superconductor and the substrate, with an increase in the proportion of current flowing through the substrate.

The reason for the decrease in current following the rise in voltage lies in the fact that for $\alpha > 1$ the resistive impedance of the superconductor increases in accordance with a $v\xi$ law, where $\xi > 1$, with rising voltage (or more precisely with rising temperature of the conductor due to the increase in Joule heat liberation in the substrate with rising voltage). If $\alpha < 1$, then $\xi < 1$, and the current in the conductor increases with rising voltage. Finally, for $\alpha = 1$ and $\xi = 1$ the resistive impedance increases linearly with rising voltage. Formally this is realized as an infinitely large resistance on the section $a−b$; while $v \leq 1$ the current in the combined conductor does not increase ($I = I_c$) although the voltage on the conductor continues rising.

The states of the system corresponding to the line $a - d$ are unstable. Actually if the combined conductor is connected to a source which ensures maintenance of a specified current the following situation is characteristic of the states corresponding to the section $a - b$. If the current in the circuit fluctuates downward by a certain small quantity, the current source automatically increases the voltage on the conductor, tending to restore the current to its original value. However, for the states under consideration, this rise in voltage will lead to a further reduction in current, and so on, until the system is in a state corresponding to the point d. In an analogous manner if the current fluctuates upward, the system will pass into the same state.

Thus for a combined conductor having $\alpha > 1$ in the current range $I_m \leq I \leq I_{ct}$ the stable states correspond to the sections $e - a$ (on raising the current from values $\iota < \iota_m$) and $b - d$ (on reducing the current from values $\iota > 1$); the section $a - d$ corresponds to the unstable states.

Let us now consider the case $\alpha < 1$ (Fig. 4-4). As in the previous cases, on increasing the current I the superconductor remains in the superconducting state until the current reaches the critical value I_{ct}. During the rise in current, for $I < I_{ct}$ the temperature of the combined conductor will be equal to the temperature of the helium tank. Let us assume that on exceeding the critical current the superconductor passes into the normal state and at the initial instant the total current $I = I_{ct}$ is displaced into the substrate. As already discussed, the equilibrium temperature of the combined conductor T will then be lower than the critical temperature T_{c0}. The superconductor will pass into the resistive state, and some of the current will return to the superconductor

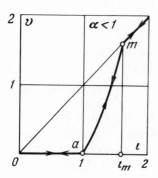

Fig. 4-4. Volt−ampere characteristic of the combined conductor for $\alpha < 1$.

from the substrate. Under these equilibrium conditions (in which the current is divided between the superconductor and the substrate), over a certain range of currents the temperature of the combined conductor will satisfy the condition $T_t < T < T_{c0}$.

For a certain current ι_m and for $\alpha < 1$, the equilibrium temperature of the conductor will reach the value T_{c0} and the total current will be expelled into the substrate. On exceeding this limiting value the total current will flow solely in the substrate and the superconductor will remain in the normal state.

The value of ι_m may be determined either from Eq. (4-17) using the condition $f = 1$ or from Eq. (4-19) using the condition $\tau = 0$. For $\alpha < 1$ we obtain $\iota_m = 1/\sqrt{\alpha}$, which naturally coincides with the value derived from Eq. (4-28) when $\alpha > 1$. The volt–ampere characteristic of such a conductor in the range $1 \le \iota \le \iota_m$ is calculated by means of Eq. (4-18); it is depicted in Fig. 4-4 (line $a-m$).

Thus a combined conductor in which $\alpha < 1$ remains superconducting for all currents $I < I_{ct}$; hence for such a conductor the effect of current degradation is completely eliminated (as in the case of a conductor having $\alpha = 1$). On exceeding the critical current the superconductor passes into the resistive and not the normal state; an increase in current then leads to a monotonic rise in the voltage across the conductor. Only for a current of $I = I_m$ does the superconductor pass into the normal state and the total current into the substrate.

It is important to emphasize that for $\alpha < 1$ the transition of the conductor through the current $\iota = 1$ is completely reversible. We see from Fig. 4-4 that after either a rise or a decrease in the current the conductor passes into exactly the same states (in contrast to the case of $\alpha > 1$ indicated in Fig. 4-3). The resistance $R = dU/dI$ is positive everywhere on the line $a-m$.

Thus the combined conductor in which $\alpha < 1$ is completely stabilized. For such a conductor, exceeding the critical current does not force the superconductor into the normal state but only causes a certain voltage to appear in the conductor, and this vanishes when the current decreases to the critical value.

It is clear from the foregoing discussion that a superconducting magnetic system made from a completely stabilized conductor is guaranteed against transitions into the normal state for any currents up to I_c (and also in the range I_{ct} to I_m). Another im-

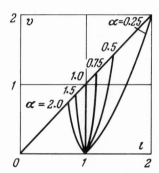

Fig. 4-5. Family of volt—ampere characteristics of the combined conductor for various values of α.

portant advantage of such a system is the fact that, since the current-degradation effect is excluded, in designing the system the characteristics of the superconductor derived from tests on short samples may be employed. In systems made from an unstabilized conductor an exact calculation of the parameters on the basis of tests on short samples is practically impossible; in the best case only empirical estimates may be used. A family of volt—ampere characteristics relating to a combined conductor is shown in Fig. 4-5 for various α values.

As already indicated a reduction in α is effected (other conditions being equal) by increasing the amount of normal metal in the combined conductor and also by increasing its purity with a view to reducing the resistivity. The value of α is also reduced by increasing the heat-transfer coefficient h from the surface of the combined conductor to the helium tank.

It is important to note that α depends quite heavily on the magnetic field induction. This is due to the following causes:

1. The critical current of the superconductor at the temperature of the helium tank I_c decreases with increasing field.
2. The critical temperature at zero current T_{c0} also decreases with increasing magnetic field.
3. The resistivity of the normal metal ρ increases with rising B (magnetoresistive effect).

On the whole α diminishes with increasing magnetic field. The α(B) relationship is shown in Fig. 4-6; the magnetic field induction at which $\alpha = 1$ is denoted by B_m. The same figure presents the curves relating to the currents I_{ct} and I_m.

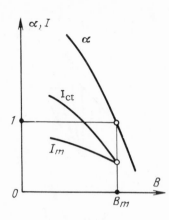

Fig. 4-6. Parameter α and currents I_{ct} and I_m as functions of magnetic induction.

For large values of B one particular combined conductor may have $\alpha < 1$ (i.e., it may be completely stabilized), while for smaller values of the induction $\alpha > 1$. It is well known that the magnetic field differs in different parts of the coil, being smallest in the outer layers. Hence, the outer layers of the winding may be incompletely stabilized. In order to eliminate this possibility it is recommended that the coil should be made in a sectional form, the outer sections being made from conductors with a higher value of α, i.e., with a greater proportion of normal metal for the same cross section of the superconductor.

As already mentioned, the system of equations (4-17) to (4-19) describing the state of thermal equilibrium of the combined conductor was obtained on the assumption that, when the current was divided between the superconductor and the substrate, a current equal to the critical current of the superconductor at the specified equilibrium temperature T flowed through the superconductor:

$$I_s = I_c(T, B). \tag{4-30}$$

When the current is divided between the superconductor and the substrate the superconductor is in fact in the resistive state. Since the substrate and the superconductor may be regarded as conductors connected in parallel, the voltage drop in the superconductor U will be equal to the voltage drop in the substrate, given (referred to unit length of the conductor) by Eq. (4-12).

It was mentioned in Section 2.4 that the potential difference U in a superconductor existing in the resistive state only developed when a current $I_s > I_c(T, B)$ was flowing through the superconductor. In accordance with (2-36) we thus have

$$U = (I_s - I_c) R_{res}, \tag{4-31}$$

where R_{res} is the resistance of the superconductor per unit length in the resistive state.

Equation (4-30) is not entirely correct since, if a current $I_c(T, B)$ did flow through the superconductor when the latter shared its current with the normal metal (substrate), there would be no voltage drop in the superconductor. Yet a nonzero potential drop is in fact preserved in the superconductor by the substrate connected in parallel to it; this was first pointed out by Sychev and Al'tov [52].

Under the conditions in question a current exceeding the critical value for the specified equilibrium temperature of the conductor T flows through the superconductor. The excess $\Delta I_s = I_s - I_c$ is determined by the obvious condition that the voltage drop in the superconductor characterized by Eq. (4-31) should be equal to the voltage drop in the substrate given by Eq. (4-12).

It follows from (4-31) that

$$I_s = I_c \,(T, \; B) + \frac{U}{R_{res}}, \tag{4-32}$$

whence from (4-12)

$$I_s = I_c \,(T, \; B) + fI \, \frac{\rho}{AR_{res}}, \tag{4-33}$$

or

$$I_s = I_c (T, \; B) + fIx, \tag{4-34}$$

where $x = \rho/AR_{res}$ is the ratio of the substrate resistance to the resistance of the superconducting part of the combined conductor in the resistive state.

From (4-11) and (4-34), we obtain [52]

$$(1-f)I = I_c(T, \; B) + fIx. \tag{4-35}$$

This equation replaces the Stekly equation (4-15) in the system of relationships describing the state of the combined conductor. The remaining equations in this system (4-12), (4-14), and (4-3) remain intact.

Using (4-3) Eq. (4-35) may be written in the form

$$(1 - f)\, \frac{I}{I_c} = 1 - \frac{T - T_t}{T_{c0} - T_t} + f \, \frac{I}{I_{ct}} \, x. \tag{4-36}$$

Equations (4-12), (4-14), and (4-36) may be made dimensionless:

$$f = \frac{\iota - 1}{\iota \, (1 - \alpha\iota + x)}; \tag{4-37}$$

$$\upsilon = \frac{\iota - 1}{1 - \alpha\iota + x}; \tag{4-38}$$

$$\tau = \frac{\alpha\iota \, (\iota - 1)}{1 - \alpha\iota + x}. \tag{4-39}$$

These equations differ from the dimensionless Stekly equations (4-17), (4-18), and (4-19) in that the ratio of the resistances x appears in the denominators.

These corrections to the Stekly model are chiefly of interest from the point of view of introducing a basic refinement to our ideas regarding the processes taking place in the combined conductor. Furthermore, as indicated in Ref. 52, in a number of cases these corrections are extremely important even in absolute magnitude, so that neglecting them may lead to considerable error in the calculations.

It should be emphasized that the model of the combined conductor considered in this paragraph does not allow for a number of effects which have considerable influence on the equilibrium conditions in the conductor. These effects amount to the following:

1. The superconductor—substrate boundary is characterized by a certain thermal contact resistance.
2. Since the thermal conductivity of the superconductor is not very great, the temperature gradient along the radius of the superconducting filament differs from zero.
3. The temperature dependence of the power carried away from the surface of the combined conductor into the helium tank is strongly nonlinear; at a certain value of the temperature difference a boiling crisis occurs.
4. The resistivity of the substrate material varies considerably with temperature for $T > T_c$.

Allowance for these factors in a number of cases leads to very substantial changes in the volt—ampere characteristic by comparison with the model which we have just described.

4.2. Influence of Contact Thermal Resistance at a Superconductor — Substrate Boundary

In any real combined conductor there is in general a specific thermal (as well as electrical) resistance at the contact surface. This resistance may be due to the presence of oxide films on the surface of the superconducting filaments, a poor quality of the contact between the superconductor and the substrate, and other causes of a technological character. The question as to the influence of thermal resistance at the superconductor—substrate interface on the equilibrium conditions in the combined conductor was considered in general terms in [53].

The values of the Joule heat evolution per unit length in a superconductor in the resistive or normal state (Q_s) and in the substrate (Q_{sub}) are given as in the case just under discussion by the equations

$$Q_s = UI(1-j); \quad Q_{sub} = UIj \qquad (4\text{-}40)$$

or considering (4-13)

$$Q_s = \frac{\rho I^2}{A} f (1 - f); \quad Q_{sub} = \frac{\rho}{A} I^2 f^2. \tag{4-41}$$

When the temperature gradient along the conductor is equal to zero, all the heat Q_s is transmitted to the substrate through the interface between the latter and the superconductor. If this surface possesses a thermal resistance, the temperature difference between the superconductor and the substrate is given by the equation

$$T_s - T_{sub} = \frac{\rho I^2}{A h_i P_i} f (1 - f), \tag{4-42}$$

where h_i is the heat-transfer coefficient from the superconductor to the substrate and P_i is the perimeter of the superconductor.

Clearly all the heat generated in the combined conductor should be transferred into the helium tank:

$$Q = Q_s + Q_{sub} = \frac{\rho}{A} I^2 f. \tag{4-43}$$

It follows that the temperature difference between the substrate and the helium tank is

$$T_{sub} - T_t = \frac{\rho I^2}{A h P} f. \tag{4-44}$$

Using (4-42) and (4-44) we obtain the following equation for the temperature difference between the superconductor and the helium tank:

$$T_s - T_t = \frac{\rho I^2}{A h P} f + \frac{\rho I^2}{A h_i P_i} f (1 - f). \tag{4-45}$$

Allowing for Eq. (4-16) we have

$$\alpha_i f^2 \iota^2 - [(\alpha_i + \alpha) \iota - 1] f \iota + 1 - \iota = 0, \tag{4-46}$$

where

$$\alpha_i = \frac{\rho I_c^2}{A h_i P_i (T_{c0} - T_t)}. \tag{4-47}$$

If the thermal resistance is absent ($h_i = \infty$, $\alpha_i = 0$), Eq. (4-46) transforms into Eq. (4-17).

Solving Eq. (4-46) for f we obtain

$$f = \frac{1}{2} \left[\frac{\alpha + \alpha_i}{\alpha_i} - \frac{1}{\alpha_i \iota} \pm \sqrt{\left(\frac{\alpha + \alpha_i}{\alpha_i} - \frac{1}{\alpha_i \iota} \right)^2 - 4 \frac{1 - \iota}{\alpha_i \iota^2}} \right]. \tag{4-48}$$

Substituting this equation into (4-12) we have

$$\upsilon = \frac{1}{2} \iota \left[\frac{\alpha + \alpha_i}{\alpha_i} - \frac{1}{\alpha_i \iota} \pm \sqrt{\left(\frac{\alpha + \alpha_i}{\alpha_i} - \frac{1}{\alpha_i \iota} \right)^2 - 4 \frac{1 - \iota}{\alpha_i \iota^2}} \right]. \tag{4-49}$$

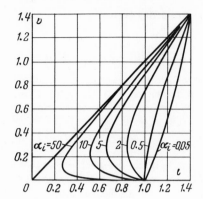

Fig. 4-7. Volt—ampere characteristics
of a combined conductor for various
values of α_i (α = 0.5).

Substituting for f by means of (4-48) in (4-45) we obtain

$$\tau = 1 - \iota + \frac{1}{2}\left[\frac{\alpha + \alpha_i}{\alpha_i} - \frac{1}{\alpha_i \iota} \pm \sqrt{\left(\frac{\alpha + \alpha_i}{\alpha_i} - \frac{1}{\alpha_i \iota}\right)^2 - 4\frac{1-\iota}{\alpha_i \iota^2}} \right]. \qquad (4\text{-}50)$$

These equations replace Eqs. (4-17) to (4-19) when allowance for the thermal
resistance at the superconductor—substrate boundary is made.

Figure 4-7 [53] shows how the line α = const is transformed in the volt—ampere
diagram for various values of α_i. Whereas for small α_i the curve corresponding to
α = 0.5 has an "ordinary" form, with increasing α_i it changes its curvature and is
displaced in the small-current direction. This means that despite the fact that $\alpha < 1$
the combined conductor is only stabilized in the range of currents not exceeding the
value of ι_m, determined by the condition

$$[\partial \upsilon / \partial \iota]_{\alpha,\, \alpha_i} = \infty.$$

Using methods analogous to those employed in Section 4.1, we obtain

$$\iota_m^2 = \frac{(\alpha - \alpha_i) \pm 2\sqrt{\alpha_i(\alpha + \alpha_i)^2 - \alpha \alpha_i}}{(\alpha + \alpha_i)^2}. \qquad (4\text{-}51)$$

For α_i = 0 this equation transforms into Eq. (4-28).

The value of ι_m may easily be determined from the known values of α and α_i
using the nomograms presented in Fig. 4-8 [53]. In this diagram the continuous lines
correspond to values of ι_m = const and the broken lines to $U_m A / \rho I_{ct}$ = const (U_m
corresponds to the point with a vertical tangent on the line α = const).

It follows from Eq. (4-49) that the slope of the line α = const at the point ι = 1,
f = 0 is given by the following equation:

$$\left[\frac{\partial (UA/\rho I_{ct})}{\partial \iota} \right]_{\substack{\iota=1, \\ f=0}} = \frac{1}{1 - (\alpha + \alpha_i)}. \qquad (4\text{-}52)$$

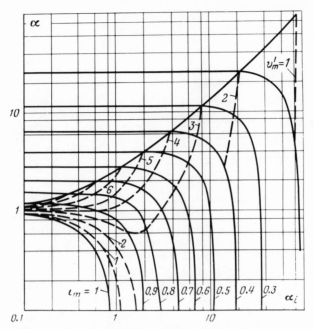

Fig. 4-8. Nomogram for determining the dimensionless current ι_m.
$$v_m = U_m A / \rho I_{ct}.$$

Thus the combined conductor will be completely stabilized for

$$\alpha + \alpha_i < 1, \tag{4-53}$$

since the derivative (4-52) will then be positive.

These effects appear most sharply in twisted superconducting cables without impregnation. Careful cleaning of the original superconducting and copper filaments, followed by impregnation of the cable with highly purified indium, greatly improves the service characteristics of the combined conductor, owing to the reduction of contact resistance at the superconductor—substrate interface. For the widely used combined conductors prepared by modern metallurgical technology, there is usually excellent bonding between the superconductor and the substrate, and the influence of contact resistance is quite negligible.

4.3. Influence of the Finite Thermal Conductivity of a Superconductor on the Stability of Combined Conductors

Up till now we have assumed that the temperature is constant over the whole cross section of the superconductor and equal to the temperature near the surface.

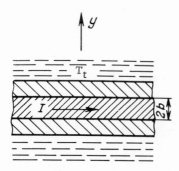

Fig. 4-9. Strip of superconducting
alloy encased in a shell (sheath).

However, when the superconductor has passed into the resistive state, heat is evolved
in the interior; the thermal conductivity being finite, this assumption therefore requires quantitative verification. It is also desirable to study the influence of the nonuniformity of the temperature distribution over the cross section of the superconductor
in cases in which this influence cannot be neglected.

In order to study this problem we analyze the following very simple case [54].
We consider a plane strip of superconducting alloy of width a and thickness $2b$ enclosed in a shell of stabilizing material of cross section A (Fig. 4-9). We shall consider that the thermal conductivity of the stabilizing shell is fairly high, and the temperature change over the shell cross section may thus be neglected. For simplicity
we assume that the thermal conductivity of the superconductor is constant and in
particular independent of the temperature. Furthermore we shall neglect the change
in temperature in directions lying in the plane of the strip. This means in particular
that in the direction of propagation of the current all these quantities vary quite
smoothly. Thus in order to study the present case it is sufficient simply to consider
the one-dimensional heat-conduction equation

$$\lambda_s \frac{\partial^2 T}{\partial y^2} + \rho_s(y) J_s^2(y) = 0, \tag{4-54}$$

where λ_s is the thermal conductivity of the superconductor, ρ_s is its resistivity, and $J_s(y)$
is the current density in the superconductor. The coordinate y is reckoned from the
median layer of the strip in a direction perpendicular to its plane.

The smooth variation of all these quantities along the current direction allows
us to consider that the voltage drop along the conductor is the same at all points of
the cross section and equal to the voltage drop in the normal metal

$$\rho_s J_s = \rho_n J_n = \rho_n I_n / A. \tag{4-55}$$

As usual the indices s and n indicate that the corresponding quantities relate to
the superconductor and the normal metal.

Since on increasing the current density in the superconductor above the critical
value its resistance increases sharply, any further increase in the current density is
practically impossible. To a first approximation we may therefore consider that at
every specified point the current density equals the critical value at the specified

temperature $J_c(T)$. We shall further consider as earlier (Section 4.1) that J_c varies linearly with temperature

$$J_c(T) = J_{ct}\left(1 - \frac{T - T_t}{T_{c0} - T_t}\right).$$
(4-56)

Allowing for the assumptions made, the original equation (5-54) takes the following form:

$$\lambda_s \frac{\partial^2 T}{\partial y^2} + \rho_n \frac{I_n}{A} J_{ct}\left(1 - \frac{T - T_t}{T_{c0} - T_t}\right) = 0.$$
(4-57)

This equation is linear. It is quite easy to obtain an exact analytical solution and fairly simple to study the properties of the solution for various values of the parameters. In accordance with (4-20) we introduce the following notation: $\iota_s = I_s/I_{ct}$ is the dimensionless current in the superconductor, $\iota_n = \iota_n/I_{ct}$ is the total current in the combined conductor, and $r = 1 - \iota_s/\iota = \iota_n/\iota$ is the dimensionless resistance of the conductor.

Using this notation Eq. (4-57) may be converted to the form

$$b^2 \frac{\partial^2 \tau}{\partial y^2} + \delta \iota r (1 - \tau) = 0,$$
(4-58)

where

$$\delta = \frac{\rho_n I_{ct}^2 b}{\lambda_s A (T_{c0} - T_t) \times 2a}.$$
(4-59)

This quantity constitutes the dimensionless thermal resistance of the whole thickness of the superconducting material. In its structure, Eq. (4-59) is similar to (4-22), defining the stabilization parameter α, which is also a dimensionless thermal resistance. It should be noted that by virtue of Eq. (4-55) the definition of the parameter δ (4-59) incorporates the combination $\rho_n I_{ct}^2 /A$ containing information relative to the properties of the normal metal. Thus the internal thermal resistance of the conductor associated with the finite thermal conductivity inside the superconductor is determined by the parameters of the normal metal as well as those of the superconductor itself.

It is interesting to carry out a numerical estimation of the values of δ for specific cases. Unfortunately existing data regarding the values of the most important parameter, the thermal conductivity of the superconductor λ_s, are extremely sparse and unreliable. Estimates based on the Wiedemann—Franz law lead to values of λ_s of the order of 10^{-2} W/(cm-K) at 4 K. The literature [55] contains values of $\lambda_s \approx 10^{-3}$ W/(cm-K) at 4.2 K for widely used alloys of the Nb—Ti system. Thus for purposes of estimation we may take a value of $\lambda_s = 3 \times 10^{-3}$ W/(cm-K).

As an example, let us consider a strip conductor 1 cm wide and 2.5 mm thick including a superconducting strip 0.55 mm thick. For a copper coating at $T \approx 4$ K subject to an induction of about 5 T, we may assume that $\rho_n = 4 \times 10^{-10}$ Ω-m. For a current density in the superconductor of $J_{ct} = 2 \times 10^9$ A/m^2 the critical current is 5×10^3 A. The critical temperature T_{c0} may be taken as 8 K. Substituting

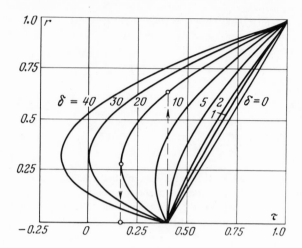

Fig. 4-10. Resistance of unit length of the conductor as a function of the dimensionless temperature τ.

all these numerical values into Eq. (4-59) we obtain $\delta \approx 4$. Thus the value of δ may greatly exceed unity and the effect under consideration will have to be taken into account when studying the stabilization conditions.

For a more detailed quantitative study we turn again to Eq. (4-58). The solution of this equation (symmetrical with respect to the middle of the strip) takes the form

$$\tau(y) = 1 - F \cosh \sqrt{\delta \iota r} \; \frac{y}{b}. \tag{4-60}$$

In order to determine the constant F in this solution we calculate the total current flowing along the superconductor for a specified temperature distribution:

$$I_s = \int_{-b}^{b} J_c(\tau) \, a \, dy = aJ_{ct} \int_{-b}^{b} (1 - \tau) \, dy = baJ_{ct}F \frac{2 \sinh \sqrt{\delta \iota r}}{\sqrt{\delta \iota r}} = FI_{ct} \frac{\sinh \sqrt{\delta \iota r}}{\sqrt{\delta \iota r}}.$$

$$\tag{4-61}$$

Thus we have

$$F = \iota_s \frac{\sqrt{\delta \iota r}}{\sinh \sqrt{\delta \iota r}} \tag{4-62}$$

and

$$\tau(y) = 1 - \iota_s \frac{\sqrt{\delta \iota r}}{\sinh \sqrt{\delta \iota r}} \cosh \sqrt{\delta \iota r} \frac{y}{b}. \tag{4-63}$$

On the external surface of the superconductor the temperature equals the specified value τ_{ext}:

$$\tau_{ext} = 1 - \iota(1 - r) \sqrt{\delta \iota r} \coth \sqrt{\delta \iota r}. \tag{4-64}$$

This equation characterizes the relationship between the resistance of unit length of the conductor r and the temperature on its surface.

Figure 4-10 shows the $r(\tau)$ relationship for a fixed current $\iota = 0.6$ and a series of increasing values of the parameter δ. In Fig. 4-11 we have the relationship between the temperature at the center of the sample and the temperature on its surface subject to the same conditions.

We see from these figures that as the parameter δ tends to zero the resistance r in the resistive region increases linearly in accordance with the data of Section 4.1, and there is no undue heating in the center. On increasing δ the resistance rises more sharply and considerable heating occurs in the center. Finally, for a certain value of δ the function $r(\tau)$ at the initial point of the resistive region acquires a vertical tangent. For still greater values of δ a slight rise in temperature above the critical value for the specified current $\tau = 1 - \iota$ should result in the abrupt appearance of a finite resistance, and at the same time finite heating in the center of the superconductor.

It follows from Fig. 4-11 that for any values of the parameter δ the maximum temperature in the center of the conductor never exceeds T_{c0} (provided of course that the external temperature remains below T_{c0}). Allowing for the assumptions made, this behavior may readily be understood: As soon as the temperature in the center rises, the resistance also increases without limit; this automatically limits the current, and hence also the heat evolution and any further temperature rise in the central region.

Let us calculate the value of δ at which the possibility of an irreversible change in the resistance r first arises. For this purpose it is sufficient to equate the derivative of the function $\tau_{ext}(r)$ given in Eq. (4-64) to zero for $r = 0$. It is convenient to use the series expansion of tanh x in this connection

$$\tanh\ x = x - \frac{1}{3} x^3 + \cdots \qquad (4\text{-}65)$$

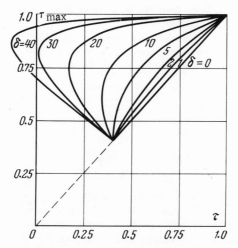

Fig. 4-11. Temperature in the center of the sample relative to that at the surface.

Close to the point $r = 0$ the function $\tau_{ext}(r)$ takes the form

$$\tau_{ext}(r) \approx 1 - \iota(1-r)\frac{1}{1-\frac{\delta\iota r}{3}} \approx 1 - \iota + \iota r - \frac{\delta\iota^2}{3}r. \qquad (4\text{-}66)$$

From

$$\partial\tau_{ext}/\partial r = 0 \qquad (4\text{-}67)$$

we obtain

$$\delta_{lim} = 3/\iota. \qquad (4\text{-}68)$$

Figures 4-10 and 4-11 also show lines decreasing in the direction of negative τ values (we remember that $\tau = 0$ corresponds to $T = T_t$). Such solutions of Eq. (4-60) show that, even for very intensive cooling of the surface, the resistive state, once having been created, may be sustained by virtue of the poor thermal conductivity in the interior of the superconductor material. The question as to the stability of such states will be considered shortly.

There is no difficulty in generalizing these results to the case of a circular wire. In so doing the hyperbolic functions are replaced by Bessel functions, while the role of the parameter δ is played [56] by

$$\delta_0 = \frac{\rho I_{ct}^2}{\pi\lambda_s(T_{c0}-T_t)A}. \qquad (4\text{-}69)$$

The dependence of δ_0 on the wire dimensions is expressed in implicit form in this equation via the dependence of I_{ct} on the cross-sectional area. The form of the $r(\tau)$ curves remains intact; however, the exact quantitative relationships are slightly different. For example, the vertical tangent on the $r(\tau)$ curve appears for a relatively large value of the parameter δ_0 equal to $8/\iota$. Clearly, in a circular conductor a greater proportion of the cross section lies closer to the surface than in a flat strip, and the cooling conditions are therefore rather more favorable.

In order to study the influence of the effect under consideration on the thermal stability of the conductor for a constant temperature along its length, it is again convenient to turn to the volt–ampere characteristics of a section of conductor. For simplicity we shall assume that heat release obeys a linear law, i.e.,

$$\tau_{ext} = \alpha r\iota^2 = \alpha\upsilon\iota. \qquad (4\text{-}70)$$

Using (4-64) we have

$$\tau_{ext} = 1 + (\upsilon-\iota)\sqrt{\delta\upsilon}\coth\sqrt{\delta\upsilon}, \qquad (4\text{-}71)$$

where

$$\iota = \frac{\upsilon\sqrt{\delta\upsilon}\coth\sqrt{\delta\upsilon}+1}{\alpha\upsilon+\sqrt{\delta\upsilon}\coth\sqrt{\delta\upsilon}}. \qquad (4\text{-}72)$$

The corresponding characteristics are presented in Figs. 4-12 and 4-13 for two cases $\alpha = 0.5$ and $\alpha = 1.5$. Comparison between these characteristics and the curves of Fig. 4-7 shows that the influence of the finite thermal conductivity of the material for the superconductor is in this case completely analogous to that of the transient thermal resistance; this is quite reasonable since both mechanisms are of a similar nature.

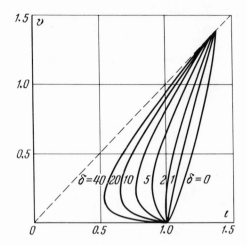

Fig. 4-12. Volt—ampere characteristics of combined conductors for $\alpha = 0.5$.

Analyzing Eq. (4-72) we easily obtain the condition under which, subject to
the condition $\alpha < 1$, the current ι_m never falls below the critical value ($\iota = 1$). For
this purpose it is again sufficient to use the first two terms of expansion (4-65). The
corresponding condition is written in the form

$$\alpha + \delta/3 < 1. \qquad (4\text{-}73)$$

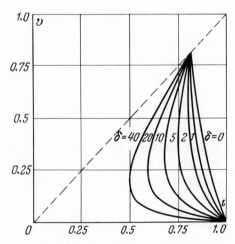

Fig. 4-13. Volt—ampere characteristics of combined conductors for $\alpha = 1.5$.

Analogously for a superconductor of circular cross section

$$\alpha + \delta_0/8 < 1. \tag{4-74}$$

It is interesting to find the value of the parameter δ for which there is no reduction of the quantity ι_m below the "ordinary" value of $\iota_m = \alpha^{-1/2}$ for the case $\alpha > 1$. If we again make use of the expansion (4-65) this value may readily be found:

$$\delta = 3\,(\sqrt{\alpha} - 1) \tag{4-75}$$

or

$$\delta_0 = 8\,(\sqrt{\alpha} - 1). \tag{4-76}$$

In contrast to the two previous stability criteria (4-73) and (4-74), these relationships are, strictly speaking, only valid for small values of δ and δ_0. However, since the series (4-65) and the analogous series for the corresponding combination of Bessel functions are alternating in sign, these equations provide a reliable estimation of the quantities δ and δ_0, and indeed offer a certain reserve.

The influence of δ on the current ι_m defined by the equation $\partial v/\partial \iota = 0$ may be studied in more detail by analyzing the exact relation of the volt–ampere characteristic (4-12). However, since the parameters of the material defining δ are usually not very reliably known, as indicated earlier, there is not a great deal of point in conducting such an analysis here. For approximate calculations it is always sufficient to use either the expansion (4-65) (for small δ) or the asymptotic representations of the corresponding functions in the opposite case. Further discussion of the influence of the internal thermal resistance of the superconductor on the equilibrium conditions in the presence of a temperature gradient is presented in Chapter 5, together with the derivation of the corresponding stability criteria.

4.4. Influence of the Boiling Crisis in Liquid Helium on the Conditions of Thermal Equilibrium in a Combined Conductor

So far we have assumed that the heat-transfer coefficient from the surface of the combined conductor in the helium tank h is constant, i.e., not varying with temperature. Actually h varies considerably with the temperature gradient, i.e., the difference between the temperatures of the combined conductor and the helium tank $\Delta T = T - T_t$. The effect of variations in the heat-transfer coefficient on the stabilization conditions was considered by Kremlev et al. [57].

The liquid helium in the tank accommodating the combined conductor is in a state of saturation (in equilibrium with its own

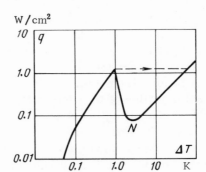

Fig. 4-14. Thermal flux density as a
function of temperature difference.

saturated vapor). Hence heat transfer from the surface of the
combined conductor to the liquid helium will under these circumstances amount to the kind of heat transfer associated with boiling.

It is well known that the boiling crisis is a characteristic of
this type of heat transfer, that is, the transition from bubble to
film-type boiling which occurs when the thermal flux density
rises above a specific level q_{cr}. Figure 4-14 shows the thermal
flux density q as a function of the temperature difference $\Delta T =
T - T_t$ for the boiling of liquid helium in a large volume [58]. We
see from this figure that the boiling crisis appears at a thermal
flux of $q_{cr} \approx 1$ W/cm^2, what corresponds to a temperature difference of $\Delta T_0 \approx 1$ K. In the case under consideration the thermal
flux is already specified; it is created by the Joule heat evolution
in the combined conductor. Under these conditions a further increase in q will lead (as indicated in the figure) to an abrupt rise
in the temperature difference ΔT, and hence (for a constant temperature of the boiling liquid) to an abrupt rise in the temperature
of the heat-emitting surface (in the present case the combined
conductor). On reducing the thermal flux the transition from film

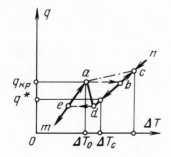

Fig. 4-15. Paths of the transition
from film to bubble-type boiling.

to bubble-type boiling will occur for a q value much smaller than q_{cr}. As indicated in Fig. 4-15, on increasing q the process takes place along the line $m - e - a - b - c - n$ and on reducing q along the line $n - c - b - d - e - m$.

It follows from Fig. 4-14 that, after the thermal flux has exceeded the value of q_{cr}, the temperature of the combined conductor will rise sharply by several tens of degrees, i.e., it will certainly exceed the critical temperature T_c. If the system maintains the current constant (inductive winding), the superconductor will pass into the normal state and all the current will be transferred to the substrate. Of course, the Joule heat evolution in the combined conductor will then suddenly rise. Whereas before the expulsion of all the current into the substrate the Joule heat evolution per unit length of the conductor was equal to $f I^2 \rho / A$ in accordance with Eq. (4-13), it will now be $I^2 \rho / A$. Thus on passing from the bubble to the film type of boiling in the liquid helium the thermal flux from the surface of the combined conductor will rise sharply (transition from point a to point c in Fig. 4-15).

We see from Fig. 4-14 that the heat-transfer coefficient also depends very heavily on temperature in the region of bubble-type boiling.

In order to allow for the temperature dependence of the heat-transfer coefficient (or more precisely the specific thermal flux) we must alter Eqs. (4-12), (4-14), (4-3), and (4-15). Instead of Eq. (4-14) we must use the more exact equation

$$qP = f I^2 \rho / A. \qquad (4-77)$$

The value of q is determined from Fig. 4-14 by reference to the quantity $\Delta T = T - T_t$.

This system of equations enables us to calculate the volt—ampere characteristic of the combined conductor with due allowance for the boiling crisis of the liquid.

Plainly in the region of states in which $T < T_t + \Delta T_0$ (here ΔT_0 is the critical temperature difference) the volt—ampere characteristic will not alter very much relative to that shown in Fig. 4-5.

We note that on the $v - \iota$ diagram plotted for the case h = const the isotherms are straight lines parallel to the $f = 1$ line. After

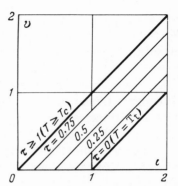

Fig. 4-16. Family of isotherms corresponding to different values of τ.

finding an expression for $\alpha(\tau, \iota)$ from Eq. (4-19) and substituting
this into (4-18) we in fact obtain

$$\frac{UA}{\rho l_c} = \tau + \iota - 1. \tag{4-78}$$

It is quite evident from this that all the isotherms corresponding to $\tau = 0-1$ are straight lines lying at an angle of $\pi/4$ to the
horizontal axis (Fig. 4-16). It follows from Eq. (4-78) that the
isotherm $\tau = 1$ (i.e., $T = T_{c0}$) coincides with the $f = 1$ line while
the isotherm $\tau = 0$ (i.e., $T = T_t$) starts at the point at which $\iota = 1$.
All the isotherms for which $\tau > 1$ coincide with the isotherm $\tau = 1$.
Of course, the isotherm $\tau = \tau_0$ (i.e., $T = T_t + \Delta T_0$) is also a straight
line; it lies between the isotherms $\tau = 1$ and $\tau = 0$.

The volt−ampere characteristic of the combined conductor,
allowing for the helium boiling crisis, appears in Fig. 4-17. We
see from this figure that the lines $\alpha = $ const experience a break

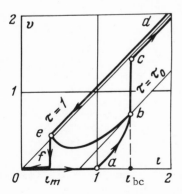

Fig. 4-17. Volt−ampere characteristic
of the combined conductor with due allowance for the helium boiling crisis.

on intersecting the isotherm $\tau = \tau_0$. Under conditions of film-type boiling the lines of the equilibrium states of the combined conductor (constituting the continuations of the lines α = const from their points of intersection with the isotherm $\tau = \tau_0$) do not, strictly speaking, correspond to the condition α = const, since the quantity h is variable. Hence the notation α = const is rather arbitrary with respect to these lines.

The lines of the equilibrium states in Fig. 4-17, of course, have the same physical meaning as those in Fig. 4-5. It is of interest to analyze the lines for which α < 1. After exceeding the critical current (ι > 1) the temperature of the combined conductor reaches τ_0, i.e., the line α < 1 intersects the isotherm (point b in Fig. 4-17), and as a result of the transition to the film type of boiling the temperature of the conductor rises sharply. The superconductor passes into the normal state and the total current passes into the substrate (line $b - c$). Plainly, on reducing the current all the current will pass through the substrate until the Joule heat evolution in the substrate decreases to a level such that the temperature difference equals $\Delta T_{c0} = T_{c0} - T_t$. The value of the specific thermal flux corresponding to this temperature head may be denoted by q* (Fig. 4-15).

It should be emphasized that for a temperature difference of ΔT_{c0} the boiling still remains of the film type. On reducing the current below ι_m the temperature of the combined conductor becomes lower than T_{c0} and the total current passes into the superconductor (line $e - f$ in Fig. 4-17).

The current ι_m may be determined by means of the following thermal-balance equation corresponding to the point e:

$$q^*P = \rho I_m^2 / A, \qquad (4\text{-}79)$$

whence

$$\iota_m = 1 / \sqrt{\alpha'}, \qquad (4\text{-}80)$$

when

$$\alpha' = \rho I_{ct}^2 / q^* P A. \qquad (4\text{-}81)$$

Structurally Eq. (4-80) is similar to (4.28)). The difference lies in the fact that instead of the Stekly stability criterion we here have the parameter α' (of a similar physical significance).

Clearly the condition for the total stabilization of the combined conductor in the case of film-type boiling may be written as

$$\iota_m = 1, \qquad (4\text{-}82)$$

whence

$$\alpha' = 1. \qquad (4\text{-}83)$$

If we formally write

$$q^* = h'(T_{c0} - T_t), \qquad (4\text{-}84)$$

where h' is the heat-transfer coefficient in the region of film-type boiling, it follows from (4-81) and (4-22) that

$$\frac{\alpha}{\alpha'} = \frac{h'}{h}. \qquad (4\text{-}85)$$

The ratio h'/h may be taken as equal to 0.03. Thus for the combined conductor to be completely stabilized with respect to the film type of boiling we must satisfy the condition $\alpha \approx 0.03$ (Fig. 4-18). In such a combined conductor the proportion of the cross section occupied by the superconductor and the constructive current density in the conductor (i.e., the ratio of the current in the superconductor to the total cross section of the combined con-

Fig. 4-18. Volt—ampere characteristic of a combined conductor completely stabilized with respect to film type of boiling.

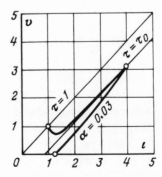

ductor) are extremely small. Hence complete stabilization with respect to the film type of boiling is not always appropriate.

When the combined conductor is completely stabilized with respect to the bubble type of boiling ($\alpha < 1$), but only partly with respect to the film type ($\alpha' > 1$, i.e., $\iota_m < 1$), the current in the combined conductor may be increased reversibly above $\iota = 1$, but not above the values corresponding to a transition from the bubble to the film type of boiling (Fig. 4-17).

When a normal zone forms in the superconductor and all the current passes into the substrate, if the current is not too great the Joule heat evolution in the substrate will not exceed q_{cr} and hence the equilibrium temperature of the conductor will not exceed τ_0. For this mode of operation the boiling of the helium will remain of the bubble type. We see that on removing the cause which produced the normal zone the superconductivity will be restored and the total current will pass back to the superconductor. The limiting current I_r for which the bubble type of boiling is not disrupted may be called the maximum superconductivity-restoration current. Thus for $I \leq I_r$ the combined conductor will be stabilized, since the possibility of a transition from bubble to film-type boiling will in this case be excluded.

In other words, by itself stabilization with respect to the film type of boiling is not essential if the possibility of a transition to the film condition is eliminated (for $\iota \leq \iota_r$). The quantity ι_r is determined from the heat-balance equation

$$q_{cr} P = \rho I_r^2 / A, \qquad (4\text{-}86)$$

whence

$$\iota_r = \frac{1}{\sqrt{\alpha_0}}, \qquad (4\text{-}87)$$

where

$$\alpha_0 = \rho I_{ct}^2 / q_{cr} P A. \qquad (4\text{-}88)$$

These equations, which are similar in structure to Eqs. (4-79) to (4-82), show that I_r does not depend on the critical current I_c.

The value of ι_r may also easily be found by using the volt–ampere characteristic. Since the thermal flux $q = IU/P$, the pro-

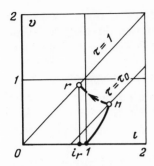

Fig. 4-19. Diagram to aid the determination of ι_τ.

duct IU should be the same at the point on the volt–ampere characteristic corresponding to the critical thermal flux q_{cr} under conditions in which the current is divided between the superconductor and the substrate (point n in Fig. 4-19) as at the point corresponding to the case in which the total current I_r is displaced into the substrate (point r). It follows that points n and r lie on the hyperbola IU/P = const. In this way we may find the values of I_r for each value of α.

Thus for $\alpha_0 \leq 1$ the combined conductor is completely stabilized

$$\iota_r = 1 \qquad (4-89)$$

for

$$\alpha_0 = 1. \qquad (4-90)$$

Figure 4-20 shows the mutual disposition on the volt–ampere characteristic of the lines α = const corresponding to complete

Fig. 4-20. Mutual disposition of the characteristic lines $\alpha = 1$, $\alpha' = 1$, and $\alpha_0 = 1$ on the volt–ampere diagram for the case in which $T_{c0} - T_t > \Delta T_0$.

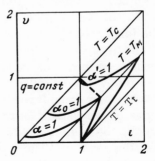

stabilization with respect to the bubble and film types of boiling $\alpha = 1$ and $\alpha' = 1$), and also with respect to the possibility of exceeding the critical thermal flux when the total current passes into the substrate ($\alpha_0 = 1$).

We see from this discussion that the whole concept of the complete stabilization of the combined conductor and the magnetic coils prepared from it undergoes a certain transformation as our knowledge of the behavior of the conductor under various working conditions deepens.

We shall now introduce a further considerable refinement into this concept. Clearly this progressive transformation of concepts may in certain cases give rise to confusion. In order to establish terminological exactitude we shall therefore in the future designate a "completely stabilized combined conductor" (and correspondingly a "completely stabilized magnetic system") to mean a conductor or system in which the possibility of an accidental transition into the normal state after the appearance of a normal-zone nucleus due to short-term actions (flux jump, motion of the coil windings) is excluded right up to the critical current.

As already indicated, the influence of the boiling crisis on the conditions of equilibrium of the combined conductors was only considered for values of the superconductor critical temperature $T_{c\,0}(B, 0)$ exceeding the minimum temperature corresponding to the onset of film-type boiling (point N in Fig. 4-14). It is nevertheless quite obvious that, since the critical temperature diminishes considerably with increasing magnetic field, the results of any calculations of the volt−ampere characteristics for the equilibrium states, in which the current is divided between the superconductor and the substrate, will differ considerably for different values of B.

We may encounter the following cases (Fig. 4-21):

1. $T_{c0}(B, 0) > T_N$, when the volt−ampere characteristics will have the form already discussed (Fig. 4-17);

2. $T_{c0}(B, 0) \leq T_M$, the superconductor passes into the normal state under conditions of bubble-type boiling (before the boiling crisis begins), and the volt−ampere characteristic has the "ordinary" Stekly form considered in Section 4.1 (Fig. 4-5). The slight difference is due to the fact that even under conditions of bubble-type boiling the heat-transfer coefficient changes slightly with changing temperature difference $T - T_t$ while in the Stekly model we assumed h = const;

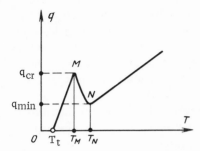

Fig. 4-21. Regions corresponding to various operating modes of the combined conductor.

3. $T_M < T_{c0}(B, 0) < T_N$, in this case the volt—ampere characteristic has the form of Fig. 4-22.

It should be noted that for a combined conductor of specified construction, i.e., a conductor with specified values of the perimeter P and substrate cross section A, specified properties of the superconductor I_c and T_c, and specified properties of the substrate material ρ for H = const, the stability criterion $\alpha = (\rho I^2)/[AhP \times (T_{c0} - T_t)]$ is only constant when the heat-transfer coefficient h is also constant. Since in actual fact h = f (T), for the lines forming the geometrical locus of the points of the equilibrium states of the combined conductor the use of the notation α = const is somewhat arbitrary. We should be careful to remember that in this case the notation in question means the line of states of a combined conductor of specified construction.

Subsequently for any combined conductor of specified construction we shall specify a value of α determined from the h value corresponding to the temperature at the onset of the boiling crisis:

$$h = \frac{q(T_M)}{T_M - T_t}. \tag{4-91}$$

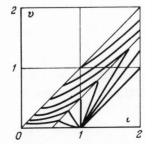

Fig. 4-22. Volt—ampere characteristics of a combined conductor at $T_M < T_{c0}(B, 0) = 6$ K.

The question as to the unstable states of a combined conductor in the temperature range $T < T_c$ requires particularly careful analysis. If the instability of particular states has not been detected in advance and the superconducting system passes into one of these states under service conditions, stability may be lost and an accident may occur.

In order to analyze the stability of the states of thermal equilibrium we may use the method of small temperature perturbations based on studying the sign of the difference

$$\frac{dq(T)}{dT} - \frac{\partial W(R, I)}{\partial T},$$ (4-92)

where $q(T)$ is the thermal flux carried out into the liquid helium from a unit area of the cooled surface, $W = \tilde{\rho}(T)I^2/AP$ is the Joule heat evolved in the conductor, referred to a unit area of cooled surface, $\tilde{\rho}$ is the resistivity of the combined conductor, and referred to a unit cross-sectional area of the substrate.

Obviously $0 \leq \tilde{\rho} \leq \rho$, where $\tilde{\rho}$ is the resistivity of the substrate material. When the superconductor is in a superconducting state, $\tilde{\rho} = 0$; for the normal state $\tilde{\rho} = \rho$. Thus $\tilde{\rho}/A$ is the effective resistance of unit length of the combined conductor.

If

$$\frac{dq(T)}{dT} - \left(\frac{\partial W}{\partial T}\right)_I > 0,$$ (4-93)

a virtual increment in the temperature will lead to a faster rise in the specific thermal flux q (faster than that of the specific power of heat evolution W). As a result of this, the combined conductor carrying the current I = const will return to the original equilibrium state. Thus in the case corresponding to condition (4-93) the state of the system will be stable.

If

$$\frac{dq(T)}{dT} - \left(\frac{\partial W}{\partial T}\right)_I < 0,$$ (4-94)

then for a virtual increment in temperature the specific power of heat evolution W will increase more rapidly than the specific thermal flux q. The heat evolved will not succeed in passing into the helium tank, and the temperature of the combined conductor ($I =$ const) will increase without limit. Obviously condition (4-94) indicates instability.

It should be emphasized that the analysis of the $q(T)$ and $W(T)$ relationships is only one of the several possible versions of this method; it may be used with equal success to consider the relationships $q(I)$ and $W(I)$ for $T =$ const, and so on. The question of stability is considered in more detail in Section 4.7.

Let us consider the stability of the conductor for various values of T_{c0}. Let us assume that $T_{c0} \leq T_M$. Considering that the heat-transfer coefficient h for the bubble type of boiling and the resistivity of the substrate material ρ are independent of temperature, we obtain

$$\frac{dq(T)}{dT} - \left(\frac{\partial W}{\partial T}\right)_I = h\left[1 - \alpha \iota^2 \left(\frac{\partial \tilde{r}}{\partial T}\right)_\iota\right], \qquad (4-95)$$

where $\tilde{r} = \tilde{\rho}/\rho$ is the dimensionless resistivity of the combined conductor (per unit area of the substrate cross section).

Assuming that the $\tilde{r}(\tau)$ relationship is linear, we have [59]

$$\tilde{r} = \frac{\tau + \iota - 1}{\iota}, \qquad (4-96)$$

whence

$$\left(\frac{\partial \tilde{r}}{\partial \tau}\right)_\iota = \frac{1}{\iota}. \qquad (4-97)$$

Thus Eq. (4-95) may be written

$$\frac{dq(T)}{dT} - \left(\frac{\partial W}{\partial T}\right)_I = h(1 - \alpha \iota). \qquad (4-98)$$

It was shown in Section 4.1 that equilibrium states of a combined conductor (in which the current was divided between the superconductor and the substrate) existed

$$\begin{array}{ll} \text{for} & \alpha > 1 \text{ when } \iota_m \leqslant \iota \leqslant 1, \\ \text{for} & \alpha < 1 \text{ when } 1 \leqslant \iota \leqslant \iota_m. \end{array} \qquad (4-99)$$

Since

$$\iota_m = 1/\sqrt{\alpha}, \qquad (4-100)$$

it follows from (4-98) and (4-99) that

$$\alpha \iota \begin{cases} \geqslant \sqrt{\alpha} & \text{for} \quad \alpha > 1; \\ \leqslant \sqrt{\alpha} & \text{for} \quad \alpha < 1. \end{cases} \tag{4-101}$$

Allowing for these inequalities we obtain from (4-95)

$$\frac{dq\,(T)}{dT} - \left(\frac{\partial W}{\partial T}\right)_I \begin{cases} < 0 & \text{for} \quad \alpha > 1; \\ > 0 & \text{for} \quad \alpha < 1. \end{cases} \tag{4-102}$$

Thus when $T_{c0} \leq T_M$, the states in which the current is divided between the superconductor and the substrate are unstable for $\alpha > 1$ and stable for $\alpha < 1$. With slight deviations which we shall subsequently indicate, this conclusion remains valid for the case in which h and ρ depend on the temperature; however, proof of the validity of this conclusion is naturally more complicated.

When $T_M < T_{c0} < T_N$, the question as to the stability of states in the temperature range between T_M and T_N may be solved extremely simply. Since in this temperature range dq(T)/dT < 0, the difference between the derivatives (4-92) is always negative, and hence for any values of α the equilibrium states in which the current is divided between the superconductor and the substrate are unstable for $T_M < T < T_N$.

In these states dU/dI > 0; it is thus clear that this condition is not always sufficient to solve the question as to the stability of the state of the combined conductor. As regards states with $T \leq T_M$, by analogy with the earlier discussion these states are unstable for $\alpha > 1$ and stable for $\alpha < 1$.

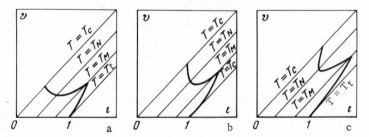

Fig. 4-23. Curves of α = const for the case in which $T_{c0} > T_N$ ($\alpha < 1$).
a) $\alpha > 0.03$; b) $\alpha \approx 0.03$; c) $\alpha < 0.03$.

Let us consider the case in which $T_{c0} > T_N$. The lines $\alpha =$ const for $\alpha < 1$ are shown in Fig. 4-23. The stability criterion determined by using the heat-transfer coefficient h' corresponding to the film mode of boiling in helium will satisfy the following conditions: $\alpha' > 1$ (Fig. 4-23a), $\alpha' \approx 1$ (Fig. 4-23b), and $\alpha' < 1$ (Fig. 4-23c).

Obviously when $T_t < T < T_M$ all the states corresponding to the division of current between the superconductor and the substrate will be unstable for $\alpha > 1$ and stable for $\alpha < 1$.

With respect to the temperature range $T_N < T < T_{c0}$, for all values of α such that $\alpha' \geq 1$ the states corresponding to the division of the current between the superconductor and the substrate are unstable. It should be emphasized that on the $v = f(\iota)$ diagram the line $\alpha =$ const for $\alpha \approx 0.03$ (i.e., $\alpha' = 1$) has a vertical tangent when $\iota = 1$.

A more detailed consideration should be given to the case $\alpha' < 1$ for $T_N < T < T_{c0}$. We see from Eqs. (4-93) and (4-94) that the point on the volt-ampere characteristic $\alpha =$ const constituting the boundary between the stable and unstable states for $T_N < T < T_{c0}$ is given by the condition

$$\frac{dq\,(T)}{dT} = \left(\frac{\partial W}{\partial T}\right)_I. \tag{4-103}$$

Considering that $W = \widetilde{\rho}I^2/AP$, for this limiting or boundary point we obtain

$$I_{\lim} = \sqrt{AP\,\frac{dq/dT}{d\widetilde{\rho}/dT}}. \tag{4-104}$$

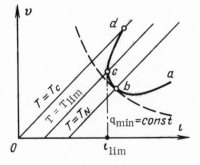

Fig. 4-24. Diagram to aid the determination of $I_{\lim}$.

TABLE 4-1. Regions of Stable and Unstable States of
a Combined Conductor When the Current Is Divided
between the Superconductor and the Substrate

Temperature	$\alpha > 1$	$\alpha < 1$
$T_{co} \leqslant T_M$	Instability	Stability
$T_M < T < T_N$	Ditto	For $T < T_M$ Stability For $T > T_M$ Instability
$T_{co} > T_N$	Ditto	For $T < T_M$ Stability For $T_M < T < T_N$ Instability $T_N < T < T_{co}$: a) For $\alpha' \geqslant 1$ Instability b) For $\alpha' < 1$: For $T_N < T < T_{lim}$ Instability For $T_{lim} < T < T_{co}$ Stability

By solving this equation together with the heat-balance equation q(T) = W(T), we may find the value of I_{lim} for a combined conductor of specified construction, i.e., with given values of A and P
(Fig. 4-24).

The foregoing conclusions as to the distribution of the regions
of stable and unstable states of the combined conductor for equilibrium states corresponding to the division of the current between
the superconductor and the substrate are presented in Table 4-1.

We see from the preceding discussions that, under current-
division conditions, the equilibrium states of the combined conductor characterized by the condition dU/dI < 0 are always unstable,
while the equilibrium states for which dU/dI > 0 are unstable in
the temperature range $T_M < T < T_N$; for any other temperatures
the states for which dU/dI > 0 are stable.

4.5. Equilibrium of a Combined

Conductor in the Normal State

We assumed earlier when considering the conditions of thermal stabilization of combined conductors that the temperature
coefficient of the resistance of the substrate material was negligibly small, i.e., the substrate resistance varied very little with
temperature. On this assumption all the isotherms corresponding
to $\tau \geq 1$ on the volt−ampere characteristic of the combined con-

ductor merge (Fig. 4-16). This means that for any temperatures $T \geq T_c$ the states of the combined conductor are stable in the thermal respect.

Actually, however, the substrate material changes its resistance considerably with temperature. By way of example Fig. 4-25 [60] shows the temperature dependence of the resistivity of copper for $\rho_{300 K}/\rho_{4.2 K} = 100$.

Theoretical analysis supported by experiment shows that a marked temperature dependence of the resistance of the substrate material exerts considerable influence on the transition of the superconducting magnetic system into the normal state [61, 62]. This compels us to take a fresh look at the reliability of many types of superconducting magnetic systems.

Let us consider how the volt–ampere characteristic of the combined conductor $UA/\rho I_c = f(I/I_c)$ transforms on making allowance for the temperature dependence of the resistance of the substrate material.

Let us assume that the superconductor has passed into the normal state (at $T = T_c$) and the whole current has been transferred to the substrate. This is clearly equivalent to the case in which an ordinary current-carrying conductor is immersed in the helium tank. If the resistance of the substrate material did not depend on the temperature, the voltage on the conductor would increase linearly with rising current. However, since Joule heat is evolved as the current increases and the temperature of the conductor rises, the $U = IR(I)$ relationship becomes nonlinear.

Thermal equilibrium is established when the temperature of the combined conductor is such that the thermal flux from its sur-

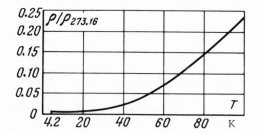

Fig. 4-25. Temperature dependence of the resistivity of copper.

face equals the Joule heat evolution. For the case under consideration, in which all the current flows through the substrate, this means that

$$q(T)P = \frac{\rho(T) I^2}{A}. \tag{4-105}$$

On the volt—ampere characteristic the geometric locus of the points corresponding to the equilibrium states of a combined conductor of specified construction (i.e., one with given A and P) may be determined for the case in which all the current appears in the substrate by means of Eq. (4-105) in the following way:

1. Using the known $\rho(T)$ relationship we plot a $W = f(I)$ curve, in which, in accordance with the earlier notation, we have $W = \rho(T)I^2/AP$ (Fig. 4-26a).
2. Using the known temperature dependence of the thermal flux associated with the boiling of liquid helium for the particular temperature T, we determine the value of q (Fig. 4-26b).
3. Using the resultant value of Q and the $W = f(I)$ curve, we determine the current I in the unknown state of equilibrium for the specified temperature T (Fig. 4-26c).
4. Using the known values of I and $\rho(T)$, we determine the voltage drop in unit length of the conductor in the specified state from the condition

$$\frac{UA}{\rho(T) I_c} = \frac{I\rho(T)}{I_{ct}\rho(T_c)}. \tag{4-106}$$

Repeating these operations for various values of T, we determine the whole set of equilibrium states of the conductor of specified construction (i.e., formally for a specified value of α = const).

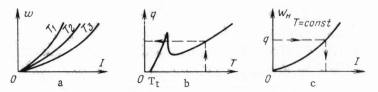

Fig. 4-26. Curves for determining the equilibrium of the states of a combined conductor when all the current flows in the substrate. $T_1 > T_2 > T_3$.

Clearly the results of such a calculation are different for different values of the critical temperature T_c, since the values of $\rho(T_c)$ are different. Furthermore when T_{c0} is small (for large fields) it may be that $T_{c0} < T_t + \Delta T_0$, where ΔT_0 is the critical temperature difference. In this case the superconductor will pass into the normal state under conditions of bubble-type boiling and the boiling crisis will occur at $T > T_{c0}$. This has considerable effect on the character of the $\alpha = \text{const}$ curves in the volt—ampere diagram.

Figure 4-27 gives the results of a calculation of the volt—ampere characteristics for various values of α ($T_{c0} = 8$ K). Let us consider the characteristic corresponding to $\alpha = 0.2$ (Fig. 4-28). The most important feature of this curve is that it lies in the region $\iota < \iota^*$, where ι^* is the current corresponding to the point d. This means that for a combined conductor of this construction there are no equilibrium states for currents $\iota > \iota^*$. In other words, if the transition of the superconductor into the normal state occurs at $\iota > \iota^*$ (for example, a transition from bubble- to film-type boiling, the $b - f$ line in Fig. 4-28) the temperature of the conductor will rise sharply and it may even melt.

For currents $\iota < \iota^*$ two possible equilibrium states may correspond to the same value of ι as indicated in Fig. 4-28 (points I and II).

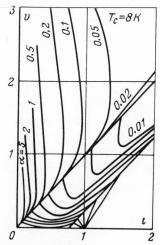

Fig. 4-27. Results of a calculation of the volt—ampere characteristics for various values of α.

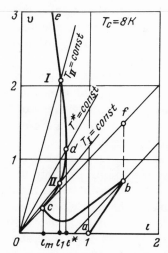

Fig. 4-28. Volt—ampere characteristic for $\alpha = 0.2$.

Clearly the equilibrium states of this combined conductor are stable for $T_{c0} < T < T^*$ (section $c - d$) since for these temperatures

$$\frac{dq\,(T)}{dT} - \left(\frac{\partial W}{\partial T}\right)_I > 0,$$

while the equilibrium states for $T > T^*$ (section $d - e$) will be unstable, since for these temperatures

$$\frac{dq\,(T)}{dT} - \left(\frac{\partial W}{\partial T}\right)_I < 0.$$

We see from Fig. 4-28 that the section $c - d$ has $dU/dI > 0$ and $d - e$ has $dU/dI < 0$.

It follows from the foregoing that the distribution of the stability and instability fields in Table 4-1 should be supplemented by the following assertion: When $T > T_{c0}$ for any values of α the combined conductor is stable for $T < T^*$ and unstable for $T > T^*$. The temperature T^* is a characteristic of the particular fabrication of the combined conductor.

The current I^* limiting the range of values for which equilibrium states may occur is one of the most important characteristics of a combined conductor. Increasing the coil current above I^* may lead to the burning out of the conductor if a normal zone appears in the superconductor and all the current is transferred to the substrate. We shall subsequently call I^* the maximum equilibrium current.

We see from the curves of Fig. 4-27 that even in cases in which $\alpha \ll 1$, i.e., for conductors stabilized with a certain reserve relative to the bubble mode of boiling in the helium, the value of I^* may be smaller than the critical current I_{ct}. For the data used in plotting this diagram ($T_c = 8$ K) the current I^* becomes equal to the critical current when $\alpha = 0.09$. Thus if for $\alpha > 0.09$ the superconductor passes into the normal state in the current range $I^* < I < I_{ct}$, there may be a subsequent unlimited rise in the temperature of the conductor.

The value of I^* is determined from the condition that the curves $q(T)$ and $W(T)$ of Fig. 4-29 should touch

$$\left(\frac{\partial W}{\partial T}\right)_I = \frac{dq\,(T)}{dT}. \tag{4-107}$$

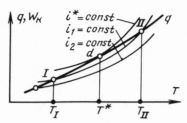

Fig. 4-29. Temperature dependence
of the thermal flux and Joule heat
evolution.

Allowing for the relationship $W = \rho(T)I^2/AP$ we obtain

$$I^* = \sqrt{AP \frac{dq(T)/dT}{d\rho(T)/dT}}. \qquad (4\text{-}108)$$

In addition to this, at the point d (as at any other point of the volt—ampere characteristic under consideration) in accordance with Eq. (4-105) the following condition is satisfied:

$$q(T) = \rho(T)I^{*2}/AP.$$

By solving the system of equations (4-107) and (4-108) with two unknowns (I^* and T) we may determine the maximum equilibrium current I^*. It follows that the current I^* is entirely determined by the construction of the combined conductor.

4.6. Volt — Ampere Characteristics

of Combined Conductors

On the basis of the foregoing arguments we may predict the behavior of combined conductors with a greater degree of reliability when these are employed in real superconducting installations.

In actual fact, for any specific combined conductor in which the thermal and electrical properties of the components are known, subjected to specified cooling conditions (free boiling, boiling in narrow channels, forced cooling), it is quite easy to construct the volt—ampere characteristics. These characteristics enable us to make a detailed analysis of the conditions of thermal equilibrium governing the conductor in a superconducting coil. By way of illustration, the Appendix (Figs. A-1 to A-9) presents the volt—ampere characteristics of combined conductors with various values of α calculated for six values of T_c (4.7-10 K). In carrying out this

TABLE 4-2

T_c, K	$\alpha_{I^*=I_{ct}}$
4.7	0.666
5.0	0.476
5.4	0.286
6.0	0.192
6.6	0.147
7.0	0.125
8.0	0.090
9.0	0.070
10.0	0.070

calculation we used existing data regarding the temperature dependence of the resistivity of copper [60], the thermal flux associated with the boiling of helium [58], the characteristics of the magnetoresistive effect in copper [51], and our own data regarding the $T_c(B, 0)$ relationship. We assumed that the character of the $\rho(T)$ relationship was the same for various values of B.

Table 4-2 presents the values of α found, in particular, by means of the nomograms in question, for which the maximum equilibrium current becomes equal to the critical current ($I^* = I_{ct}$). The relationship $\alpha_{I^* = I_{ct}} = f[T_{c0}(B, 0)]$ is illustrated in Fig. 4-30. The values of $\alpha_{I^* = I_{ct}}$ were found by means of an $\alpha = f(\iota)$ diagram on which the $\alpha_{I = I^*}$ were recorded for each temperature. The point of intersection of a particular isotherm (smooth curve linking the values of $\alpha_{I=I^*}$ at various points) with the vertical $\iota = 1$ determines the value of $\alpha_{I^* = I_{ct}}$ at the particular temperature.

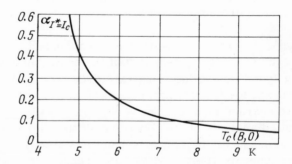

Fig. 4-30. Dependence of $\alpha_{I^* = I_{ct}}$ on the critical temperature.

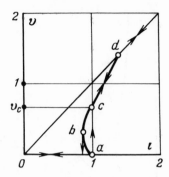

Fig. 4-31. Volt—ampere characteristic
in the region of bubble-type boiling for
$\alpha < 1$.

Let us consider the volt—ampere characteristics presented in Figs. A-1 to A-6.
We should specially notice the character of the α = const curves in the region of
bubble boiling. As already noted, the heat-transfer coefficient in this region varies
slightly with the temperature difference. Thus the real volt—ampere characteristic
differs from the Stekly characteristic constructed on the assumption that h = const.
In particular in the present case the line α = 1 has a marked curvature. For certain
values of α slightly smaller than unity, the volt—ampere characteristic in the region
of bubble-type boiling has a peculiar form, as indicated in Fig. 4-31. Of course,
the states corresponding to the section $a-b$ of this characteristic are unstable; in this
region the combined conductor behaves as partly stabilized. On exceeding the critical current a voltage suddenly appears in the conductor. The transition from the
superconducting to the normal state is irreversible in this case.

As we remarked earlier, in the case of heat transfer to liquid helium boiling
in a large volume the temperatures T_M and T_N are approximately 5.1 and 6.7 K.
Thus the nomogram for T_{C0} = 4.7 K (Fig. A-1) corresponds to the case in which
$T_{C0} < T_M$; the nomogram for T_{C0} = 5 K (Fig. A-2) corresponds to the case in which
$T_{C0} \approx T_M$, the nomogram for T_{C0} = 6 K (Fig. A-4) to the case $T_M < T_{C0} < T_N$, and
finally the nomogram for T_{C0} = 6.6 K (Fig. A-5) to the case $T_{C0} \approx T_N$. The special
features of the volt—ampere characteristics corresponding to these cases in the temperature range $T \leq T_{C0}$ were considered in the foregoing discussions. In the temperature range $T > T_N$ for the cases under consideration the curves α = const have
no singular features distinguishing them from the curves for which $T_{C0} > T_N$.

The question as to the character of the lines α = const on the volt—ampere
characteristics in the temperature range $T_{C0} \leq T \leq T_N$ requires additional consideration for the case in which $T_{C0} < T_M$ and $T_M < T_{C0} < T_N$. These cases are illustrated
schematically in Figs. 4-32 and 4-33.*

If $T_{C0} < T_M$, then after the superconductor has passed into the normal state
(point a in Fig. 4-32) a rise in temperature takes place first in the region of bubble-

---

*These figures give the volt—ampere characteristics for combined conductors with
$\alpha < 1$. It is nevertheless quite plain that in the temperature range T_{C0} to T_N the character of the curves α = const will be qualitatively the same for any values of α.

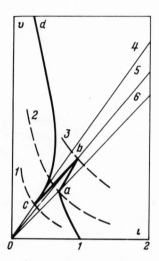

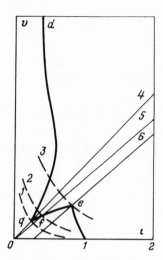

Fig. 4-32. Volt—ampere characteristic for $T_{c0} < T_M$. 1) q_{min} = const; 2) $q(T_c)$ = const; 3) q_{cr} = const; 4) T_N = const; 5) T_M = const; 6) T_c = const.

Fig. 4-33. Volt—ampere characteristic for $T_{c0} > T_M$. 1) q_{min} = const; 2) $q(T_c)$ = const; 3) q_{cr} = const; 4) T_N = const; 5) T_M = const; 6) T_c = const.

type boiling up to the point b corresponding to the temperature T_M and the thermal flux q_{cr}. On further raising the temperature, the line of equilibrium states intersects the isotherm T_N = const at the point c corresponding to the thermal flux q_{min}. For temperatures greater than T_N the curve α = const has the ordinary character (line $c-d$). Of course states corresponding to the section $a-b$ are stable and those corresponding to the section $b-c$ unstable.

When $T_M < T_{c0} < T_N$ the superconductor passes into the normal state at the point f (Fig. 4-33), and on further raising the temperature the thermal flux in the equilibrium state decreases until a state at temperature T_N has been reached (point q). In the temperature range $T > T_N$ the curve α = const has its ordinary character. It is clear that the states corresponding to section $f-q$ are unstable.

It should be noted that the volt—ampere characteristics in Figs. 4-32 and 4-33 have a qualitative character and are not shown to scale. In the real diagrams executed on the ordinary scale the isotherms T_M and T_N practically merge (Figs. A-1 and A-3).

4.7. Method of the Low-Resistance Shunt

In the preceding section we derived the detailed volt—ampere characteristics of combined conductors for a wide spectrum of critical temperatures, enabling us to present all possible situations arising during the operation of superconducting installations in explicit form.

It should nevertheless be noted that it is by no means obvious in advance now much the volt−ampere characteristics obtained by calculation will correspond to the real characteristics. Published experimental data [63-65] only refer to sections of the volt − ampere characteristics corresponding to stable states of thermal equilbrium ($\partial U/dI > 0$). In addition to this, in a number of the experiments in question there was no control over the isothermal state of the part of the combined conductor studied, and hence at least some of these data relate to cases in which the resistive zone changed in dimensions along the length of the conductor.

The absence of complete experimental data regarding the volt−ampere characteristics of combined conductors in equilibrium states (especially unstable states) is due to specific difficulties which arise in conducting such experiments. These difficulties amount to the following.

Let us consider the realization of an unstable part on the volt − ampere characteristic of a sample of combined conductor connected to a supply source. The characteristic of such a conductor appears in Fig. 4-34. Let us assume that the only reason for the superconductor passing into the normal state is that of exceeding the critical current. We shall gradually increase the current in the circuit by increasing the voltage of the supply source. On reaching the critical current the test sample acquires a resistance $R_{c.c.}$ (c.c. = combined conductor). Such a state of the combined conductor (current $I = I_c$, temperature $T = T_t$) is represented by point a in Fig. 4-34. In this state

$$R_{c.c.} = \left(\frac{\partial U}{\partial I}\right)_{\alpha, T_t} < 0. \qquad (4\text{-}109)$$

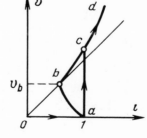

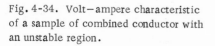

Fig. 4-34. Volt−ampere characteristic of a sample of combined conductor with an unstable region.

Strictly speaking, when the resistance $R_{c.c.}$ appears, the current in the circuit should fall. Since $R_{c.c.} < 0$ this should lead to a change of state in the conductor along the line $a - b$. Actually, however, the resistance $R_{c.c.}$ of the combined conductor in the resistive state is so small compared with the resistance of the rest of the circuit that on using practical supply sources the appearance of the resistance $R_{c.c.}$ will not lead to any decrease in the current. The current in the circuit will remain constant; the temperature of the combined conductor will rise as a result of the evolution of Joule heat in the resistance $R_{c.c.}$. The temperature will rise until the conductor reaches a state of stable equilibrium for the specified values of I_{ct} and α (point c); in this state all the current flows through the substrate. Further raising the current corresponds to states of the conductor represented by the $c - d$ section of the volt−ampere characteristic.

Thus the use of ordinary supply sources will not enable us to realize the state of unstable equilibrium of the combined conductor by direct experiment.

These difficulties may be overcome by the method of the low-resistance adjustable shunt which we proposed [66]. This method enables the properties of the supply source to be varied. A low-resistance shunt 2 composed of normal metal is connected to the source circuit in parallel with the test sample of the combined conductor 1 (Fig. 4-35). Let us denote the current in the sample by $I_{c.c.}$, the current in the shunt by I_{sh}, and the total current in the circuit by I_{tot}. Naturally

$$I_{tot} = I_{c.c.} + I_{sh} \qquad (4\text{--}110)$$

While the test sample is in the superconducting state, current

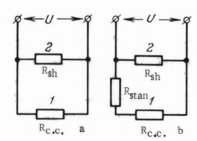

Fig. 4-35. Equivalent circuit realizing the unstable states of a combined conductor. a) Without any standard resistance; b) with a standard resistance; 1) conductor sample; 2) low-resistance shunt.

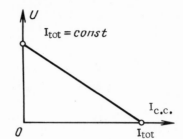

Fig. 4-36. Volt—ampere characteristic of the shunt.

does not appear in the shunt, and the voltage drops in the sample
and shunt $U_{c.c.}$ and U_{sh} are equal to zero.

After a resistance $R_{c.c.}$ has appeared in the test sample, the
total current (which hardly changes at all because of the small
value of $R_{c.c.}$) is divided between the sample and the shunt. The
voltage drops in the shunt and sample are then equal:

$$U = I_{c.c.} R_{c.c.} = I_{sh} R_{sh}. \qquad (4\text{-}111)$$

It follows from (4-110) and (4-111) that

$$U = (I_{tot} - I_{c.c.}) R_{sh}. \qquad (4\text{-}112)$$

This equation determines the volt—ampere characteristic of
the shunt when the current is divided between the sample and the
shunt, for specified values of the total current I_{tot} and shunt resistance R_{sh} (Fig. 4-36).

It is obvious that when the current is divided between these
conductors the state of the test sample with the shunt in parallel
is determined by the point of intersection of the volt—ampere char-

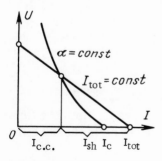

Fig. 4-37. Volt—ampere characteristics of the sample and shunt.

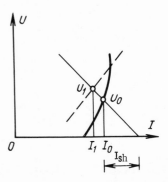

Fig. 4-38. Displacement of the point of
intersection of the volt−ampere characteristics on raising the initial temperature
of the sample.

acteristics of the sample and shunt in the $U - I_{c.c.}$ diagram (Fig.
4-37). These characteristics can only intersect when the resistance of the shunt is low. In this case the volt−ampere characteristic of the shunt is fairly oblique and intersects the sample
characteristic.

By experimentally recording the values of U and $I_{c.c.}$ we
may obtain the volt−ampere characteristic of the combined conductor.

This is essentially the method used for studying the equilibrium states of a combined conductor with the help of an adjustable low-resistance shunt.

The question naturally arises as to whether all the equilibrium
states of combined conductors may be realized in this way.

It is easy to see (Fig. 4-37) that for any quasistatic changes
in the state of the sample the point of intersection of the sample
and shunt characteristics may only be displaced from the original
position along the volt−ampere characteristic of the shunt. Figure 4-38 shows the result of such a displacement taking place, for
example, as a result of fluctuational rise in the initial sample
temperature T_0 by an amount ΔT constant along its length. The
working point is then moved to a new position (U_1, I_1).

Let us consider the heat-balance equation of the conductor
and, as before, neglect the change in temperature along its length:

$$cm \frac{\partial T}{\partial t} = RI^2 + Pq(T - T_t). \qquad (4\text{-}113)$$

Here $Pq(T - T_t)$ is the thermal flux passing into the liquid
helium; RI^2 is the power of the Joule losses; c is the specific heat;
m is the mass of a unit length of the conductor.

If we substitute the new temperature value $(T_t + \Delta T)$ into this basic equation and remember that T_t is the steady-state temperature, we obtain

$$cm \, \frac{\partial \Delta T}{\partial t} = \frac{d}{dT} \, (RI^2 - Pq) \, \Delta T, \qquad (4\text{-}114)$$

whence

$$\Delta T = \Delta T \, (0) \exp \left[\frac{1}{cm} \frac{d}{dT} \, (RI^2 - Pq) \right] t. \qquad (4\text{-}115)$$

Thus in the linear approximation ΔT rises without limit for a positive derivative $d(RI^2 - Pq)/dt$ (unstable equilibrium) and decreases for a negative gradient.

The sign of the derivative in the exponential may easily be determined directly from Fig. 4-38. The power of the losses at the new point (U_1, I_1) is determined by the product $U_1 I_1$ only, while the new value of the heat outflow is determined by the power at the point of intersection between the isotherm passing through the point (U_1, I_1) and the volt—ampere characteristic. The heat outflow q depends solely on the temperature and becomes equal to the heat evolution on the volt—ampere characteristic connecting all the equilibrium points.

It is not difficult to verify that the isotherm depicted in Fig. 4-38 corresponds to a positive value of ΔT and that in this case the equilibrium is stable. On moving along the volt—ampere characteristic the stability is interrupted after passing through the point at which the load line touches the volt—ampere curve.

Having established such a simple stability criterion, it is easy to use the diagram further in order to find the range of stability for a specified load line (a specified value of R_{sh}), and also the limiting attainable regions of stability realized when R_{sh} tends to zero. It may easily be checked, in particular, that in the region corresponding to a decrease in thermal flux with rising temperature it is impossible to attain stable equilibrium for any arbitrarily small values of R_{sh}, although the slope of the characteristics in this region may be positive.

It is clear from this discussion that the low-resistance shunt method enables us to realize a wide variety of states of the combined conductor, including those with a negative resistance

$(dU/dI < 0)$. An analogous method was first used by Stekly in order
to obtain the volt—ampere characteristics of a close-packed coil
wound with a single wire [67]. Since in the course of the experiment the resistance of the coil reached considerable values, the
method did not have to be extremely sensitive, and the shunt was
placed outside the cryostat. Attempts at using this method in order
to determine the volt—ampere characteristics of combined conductors [68] were unsuccessful owing to the extremely low resistance
of the conductor in the resistance state and the finite resistance
of the connecting leads. In our own paper [66] the sensitivity of
this method was greatly enhanced by placing the test sample, the
shunt, and the leads connecting them in a liquid-helium tank, which
minimized the resistance of the connecting leads.

In addition to this, for a direct determination of the current
flowing through the test sample, a standard resistance R_{stan} was
connected in series with the sample, its resistance being known
to a high accuracy (Fig. 4-35). In this case in order to obtain the
volt—ampere characteristic the voltage had to be recorded directly
from the combined conductor.

4.8. Experimental Results

Let us consider the results of some experiments on determining the volt—ampere characteristics of combined conductors
under isothermal conditons [69]. The method of the low-resistance
shunt described in the preceding section was realized in the following manner in this case.

The experimental apparatus depicted schematically in Fig.
4-39 was a silvered cylindrical copper block 18 mm in diameter
and 140 mm long. A slot was made in one generator of the block
along almost the entire length, and in one half a cut was made over
the cross section. In this way the block was divided into two parts,
(1, 2). An ebonite spacer (3) screwed to both parts was placed in
the slot. The two parts of the block were electrically insulated
from one another. The test sample of the combined conductor (4)
was placed between parts 1 and 2 furnished with a series of potential terminals separated at a distances of 10 mm from one another.
This sample was wound on a Teflon sprocket (5) with an external
diameter of about 25 mm.

Thus the sample, having a length of approximately 100-150 mm,
only touched the sprocket framework at individual points (at the

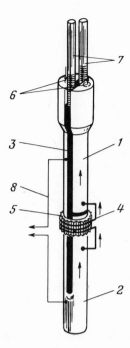

Fig. 4-39. Experimental unit for determining the volt—ampere characteristics
of a combined conductor.

tips of the teeth) and hence most of the heat evolved from the sample surface passed into the large volume of liquid helium. Current
leads (6) were attached to the two parts of the test sample at the
upper end, these being made from copper tubes. Threaded channels were made in both parts of the sample at the top end, and 30-
mm-long pins (7) were screwed into these; the pins connected parts
1 and 2 of the copper block electrically and formed a shunt. By
rotating the pins to a greater or lesser extent the resistance of
the shunt could be varied over a wide range, and this made it possible to study a number of different parts of the characteristics in
a single experiment.

Potential leads (8) were soldered to section 1 of the copper
block at distances of 100 mm from one another along the generator;
the parts of the block between these leads acted as standard resistance. This resistance was calibrated under the same conditions
as those governing the experiment itself.

The experimental device so described was placed inside a
superconducting solenoid (with an independent supply), which acted
as the source of external magnetic field. The experiment was
carried out in the following way. The current in the circuit of the

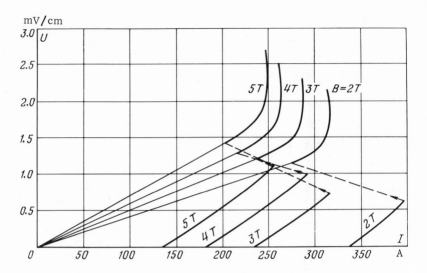

Fig. 4-40. Volt—ampere characteristics of an eight-strand conductor without insulation.

test device was gradually increased, and on reaching I_{cr} the sample under examination acquired a resistive zone, which traveled rapidly along the whole sample. After this zone had traveled along the sample, the voltage drop in the auxiliary sections neighboring the main (central) section became equal to the voltage drop in the main section. This indicated that the temperatures of the test sample were identical in these sections (to an accuracy of 0.005–0.006 K) and hence the main section was subject to isothermal conditions.* After this the principal measurements began — the signals from the principal measuring section and from the standard resistance were passed to a two-coordinate automatic recorder. By gradually increasing the total current in the circuit we were able to obtain the volt—ampere characteristic of the combined conductor under test on the automatic recorder for any specified value of the external magnetic field.

Figure 4-40 shows the experimental volt—ampere characteristics of an eight-strand conductor without insulation prepared by metallurgical technology (external diameter 1.0 mm, diameter of

---

* Additional equalizing of the temperature may also be achieved by using heaters close to the ends of the test sample.

the superconducting strand approximately 0.17 mm). The lower
(continuous) part of the line B = const corresponds to the region
of bubble-type boiling; it starts at the point $T = T_t$, $I = I_{c0}(T_t, B)$
and ends at a point corresponding to the temperature of the onset
of the boiling crisis $(T = T_M)$. The upper (continuous) part of the
line corresponds to states of the combined conductor in which
the superconductor is in the normal state and the total current
flows through the substrate $(T \geq T_{c0})$. The lower point on this
part of the line corresponds to the critical temperature of the
superconductor in the specified magnetic field at zero current.
Hence the thin continuous lines connecting these points on various
B = const characteristics to the origin of coordinates constitute
isotherms corresponding to $T \approx T_{c0}(B, 0)$.

On analyzing the B = const lines in the region of bubble-type
boiling, we may conclude that in the region of the magnetic fields
studied $(B \leq 5$ T) the eight-strand conductor has a value of $\alpha < 1$,
i.e., it is completely stabilized relative to the bubble type of boiling in a large of volume of helium.

Let us now consider the degree of stabilization of this conductor with respect to the film type of boiling. Let us assume that
as current diminishes, the conductor passes from the state corresponding to a temperature $T > T_c$ into the superconducting state.
Let us denote the current corresponding to this transition by $\tilde{I}_p$.

Figure 4-41 shows the values of the critical current $I_{ct}(T_t, B)$
and the current $\tilde{I}_p$ (curves 1 and 2) for which the superconducting

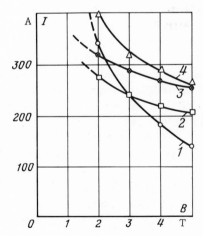

Fig. 4-41. Dependence of certain characteristic currents on the magnetic field induction for an eight-strand cable without
insulation. 1) $I_{ct}(T_t, B)$; 2) $\tilde{I}_p$; 3) $I*$; 4)
$I_{b.c.}$.

properties of the conductor are restored, taken from the volt—
ampere characteristics of Fig. 4-40.

It should be noted that, as it is not always experimentally
possible to ensure a uniform temperature distribution along the
entire length of the test sample, the restoration of the superconducting properties takes place, not at the minimum current corresponding to the existence of the normal zone I_m, but at a current
close to the minimum current of normal-zone propagation I_p (this
will be considered in more detail in Section 5.6).

We see from Fig. 4-41 that for magnetic fields greater than
about 3 T the eight-strand conductor is completely stabilized even
with respect to the film type of helium boiling in a large volume
$(\tilde{I}_p > I_{ct})$.

At temperatures $T > T_c$ (Fig. 4-40) the volt—ampere characteristics have exactly the same character as that which we established theoretically earlier. Corresponding to each value of the
induction B there is a specific value of the maximum equilibrium
current I*. The values of I* based on the data of Fig. 4-40 are
shown as functions of B in Fig. 4-41 (curve 3). It follows from
this figure that $I^* > I_{ct}$ for B > 2.2 T. Thus the conductor under
consideration is completely guaranteed against burning out in this
range of fields, should the current be transferred to the substrate
for any currents $I \leq I_{ct}$ (T_t, B).

Figure 4-41 also shows the value of the current $I_{b.c.}$ corresponding to the onset of the boiling crisis (curve 4). Over the
whole range of magnetic fields studied $I^* < I_{b.c.}$, i.e., in these
fields for $I > I_{ct}$ the boiling crisis always leads to the burning-out
of the conductor.

Apart from these samples, in order to study the effect of cooling conditions on the characteristics of the equilibrium state, we
also investigated an eight-strand conductor with organic insulation
(polyurethane paint 0.015 mm thick). The volt—ampere characteristics and the principal parameters of this conductor are shown
in Figs. 4-42 and 4-43.

We may draw the following conclusions from these figures.

1. For B < 4.2 T we have $\alpha > 1$, i.e., the combined conductor
under consideration is not stabilized in these fields even with respect to bubble-type boiling. The sharp decrease in the degree of
stabilization is due to the presence of the insulation, which has
a considerable thermal resistance. The heat-transfer coefficient

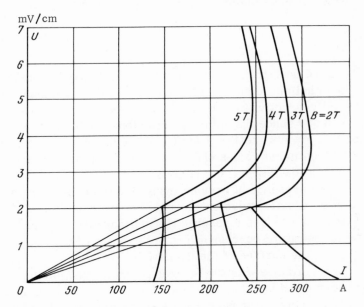

Fig. 4-42. Volt−ampere characteristics of an eight-strand conductor
with organic insulation.

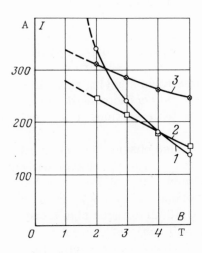

Fig. 4-43. Certain characteristics of the
currents for an insulated eight-strand conductor as functions of the magnetic field
induction. 1) $I_{ct}(T_t, B)$; 2) $\tilde{I}_p$; 3) I*.

h_i from the combined conductor to the liquid helium through the layer of insulation is determined by the well-known equation

$$h_i = \frac{1}{\frac{1}{h} + \frac{\delta}{\lambda}}, \tag{4-116}$$

where h is the heat-transfer coefficient from the cooled surface to the liquid helium; δ and λ are respectively the thickness and the thermal conductivity of the insulation. Obviously we always have $h_i < h$ and hence the stability criterion α of the insulated conductor is always greater than that of the uninsulated conductor. The difference in the values of α is only substantial when the insulation has a low thermal conductivity.

2. In the presence of organic insulation the conductor passes into the normal state under bubble-type boiling conditions. Thus in this case the conductor reaches the critical temperature earlier than the temperature corresponding to the onset of the boiling crisis T_M on the surface of the insulation.

3. In accordance with the foregoing arguments, for the conductor under consideration in a field of B < 4.2 T we have $\tilde{I}_p < I_{ct}$, the difference between these currents increasing rapidly with diminishing B; for B = 2 T we have $I_{ct} - \tilde{I}_p = 98$ A.

4. The maximum equilibrium current I* is only smaller than the critical current in fields B < 2.3 T, i.e., in practically the same range of fields as a noninsulated conductor. For larger values of B the expulsion of the current $I \leq I_{ct}$ into the substrate presents no danger from the point of view of the possible burn-out of the conductor.

It should be emphasized that it was when studying the volt-ampere characteristics of an eight-strand conductor with organic insulation that we first realized the unstable sections in the region of bubble-type boiling.

An interesting characteristic was also observed when studying the volt-ampere characteristic of a four-strand conductor without any insulation (external diameter 0.8 mm, ratio of the cross sections of the superconductor and the normal metal 0.85). In the region of bubble-type boiling (Fig. 4-44) dU/dI changes sign along the line B = const, and in a certain small range of current $I \leq I_c$ there are two equilibrium states of the combined conductor corresponding to the same value of I for bubble-type boiling.

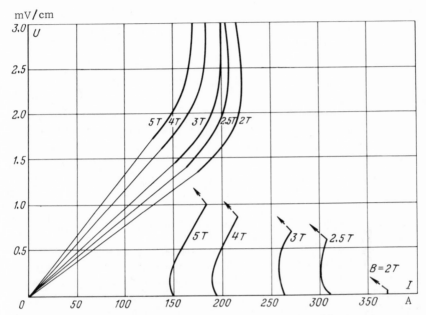

Fig. 4-44. Volt—ampere characteristic of a four-strand combined conductor.

These experimental results confirm our earlier theoretical
conclusions regarding the singularities of the volt—ampere characteristics in the region of bubble-type boiling, due to the temperature dependence of the heat-transfer coefficient in this region.
The form of the experimental volt—ampere characteristic agrees
closely with that of the characteristic in Fig. 4-31.

When analyzing the current distribution in combined conductors in Section 4.1,
we showed that whenever the current was divided between the superconductor and the
substrate a current exceeding the critical current of the superconductor for the particular equilibrium temperature T flowed through the superconductor. Allowance for
this correction led to certain relationships differing from the well-known Stekly equations in having an extra term x in the denominators of the right-hand sides of Eqs.
(4-37) to (4-39).

However, since a number of factors occur simultaneously in real combined conductors (nonlinearity of the heat-transfer coefficient h, existence of thermal and electrical resistances at the boundary between the superconductor and the substrate, nonlinearity of the volt—ampere characteristic of the superconductor in the resistive
state for T = const, and so on), and allowance for these in any theoretical analysis is

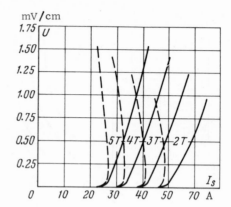

Fig. 4-45. Volt—ampere characteristics of a seven-strand combined
conductor.

very difficult, the question as to the degree of influence of the parameter x on the
current distribution in combined conductors remains open.

We shall now consider the results of an experimental investigation into the current distribution in technical combined conductors [70]. In Fig. 4-45 the continuous
lines indicate the volt—ampere characteristics of a seven-strand combined conductor obtained for various values of the external magnetic field under bubble-type boiling conditions. The conductor is 0.3 mm in external diameter and comprises a copper
matrix with seven strands of Nb—Ti alloy 0.07-0.10 mm in diameter; it is fabricated
by a metallurgical technology. The upper points of the lines B = const on the volt—
ampere characteristics correspond to the instant at which the boiling crisis begins.
Further raising the current in the sample led to an uncontrolled transition of the combined conductor into the normal state (not shown in Fig. 4-45), when all the current
flowed through the substrate ($T \geq T_c$).

Since the resistance of the superconductor in the normal state is several orders
of magnitude higher than the resistance of copper, and the resistance of copper below
10 K varies only slightly with temperature, we may consider that the resistance of the
copper matrix is

$$R_{sub} = \rho/A \approx \left(\frac{dU}{dI} \right)_{T=T_c}. \tag{4-117}$$

The proportion of current flowing through the matrix is

$$i_{sub} = fI = U/R_{sub} \tag{4-118}$$

Let us now find the current flowing through the superconductor when the latter
is in a resistive state:

$$I_s = (1-f)I = I - i_{sub} = I - U/R_{sub} \tag{4-119}$$

In Fig. 4-45 the broken lines show the $U(I_s)$ relationships corresponding to the
experimentally obtained volt—ampere characteristics of the combined conductor.

The thermal flux carried away from a unit surface of the combined conductor in
liquid helium is determined by the equation

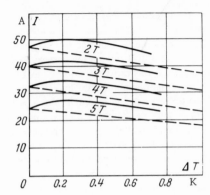

Fig. 4-46. Temperature dependences
of the current I_s (continuous lines) and
the current I_{cr} (broken lines).

$$q = \frac{UI}{Pl} = h(T)[T_{sub} - T_t], \qquad (4\text{-}120)$$

where $P = \pi d$ is the cooled perimeter of the conductor, $l = 1$ cm is the length of the
conductor under study, and T_{sub} is the temperature of the conductor (substrate).

By using one or other of the $h(T)$ relationships we may ascribe to each point on
the volt—ampere characteristic a corresponding temperature of the conductor. Since
the superconducting strands are fine and well-bonded to the copper matrix, we may
consider that $T_{sub} = T_s$.

In Fig. 4-46 the continuous lines indicate the relationships $I_s = f(\Delta T)$, where
$\Delta T = T_s - T_t$, plotted with the aid of the heat-transfer curve $q(\Delta T)$ characteristic
of the cooling conditions in the particular experiment [71]. The corresponding temperature dependences of the critical current $I_c(B, T)$ were obtained in supplementary
experiments and are shown in Fig. 4-46 by broken lines. A sample temperature exceeding the temperature of the helium tank T_t was achieved by the "inverted dewar"
method and recorded by means of resistance thermometers. From the resultant data
it is easy to find the ratio $(I_s - I_c)/I_c$ for various values of the external magnetic field.

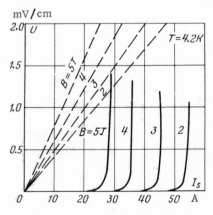

Fig. 4-47. Volt—ampere characteristics
of the substrate (broken lines) and the
superconductor in the resistive state at
$T = T_t$ (continuous lines).

We remember that in the Stekly model the current flowing through the super-conductor was taken as equal to the critical current of the superconductor for the particular temperature. Thus the ratio $(I_s - I_c)/I_c$ shows by how much the current in the superconductor differs from the current predicted by the Stekly model. We see from Fig. 4-46 that this ratio may amount to 10-15%, which is extremely important when estimating the degree of stability of superconducting devices. Such a considerable difference between the current in the superconductor and the critical current may be explained with the help of the curves presented in Fig. 4-47. In this figure the broken lines represent the volt—ampere characteristics of the substrate plotted on the basis of the $U_{sub} = IR_{sub}$ relationship, while the continuous lines represent the volt—ampere characteristics of the superconductor in the resistive state at $T = T_t$. The $U(I_s)_{T = T_t}$ relationships are found by converting the $U(I_s)$ relationships in Fig. 4-45. The conversion is effected by using the equation of [72]:

$$I_s(B,\ U,\ T) = I_s(B,\ U,\ T_t)\ [1-\tau].\qquad\qquad(4\text{-}121)$$

It is very clear from Fig. 4-47 that on the initial sections of the $U(I_s)$ isotherms the resistance of the superconductor in the resistive state R_{res} is considerably lower than the resistance of the substrate R_{sub}. Thus in this range of states of the superconductor the parameter $x = R_{sub}/R_{res}$ in Eqs. (4-37) to (4-39) is commensurable with the other quantities in this equation. This leads in particular to a considerable difference in the current I_s relative to the value based on the Stekly model, and to a change in the remaining characteristic parameters (f, v, τ) describing the state of the combined conductor when the current is divided between the superconductor and the substrate.

Chapter 5

Equilibrium of the Normal Zone in Combined Conductors in the Presence of a Longitudinal Temperature Gradient

5.1. General Principles

The model of a combined conductor set out in detail in the preceding chapter enables us to study the equilibrium of the normal zone in a conductor for the case in which the temperature is identical along its entire length, i.e., when the normal (or resistive) zone occupies the entire length of the conductor. The results so obtained are of great practical significance. However, since the sources of heat evolution in the winding are mostly of the localized type, an investigation of this kind fails to reveal every characteristic of the propagation or contraction of the normal zone along the superconductor.

If a temperature gradient exists, the boundary of the normal zone moves on account of the loss of heat from the zone in question, in which Joule heat is evolved when a current flows through the conductor. This loss of heat takes place through the boundary between the zone and that part of the superconductor still in a superconducting state. Thus the case in which the normal zone moves along the superconductor differs from the case of constant temperature along its entire length by virtue of an outflow of heat from the normal zone through the boundary and along the combined conductor.

We indicated in Chapter 4 that the normal zone could only be propagated for currents exceeding the minimum normal-zone prop-

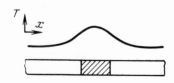

Fig. 5-1. Distribution of temperature along the combined conductor.

agation current I_p. Clearly the greater the current in the conductor by comparison with I_p the higher the rate of propagation of the
normal zone along the conductor, since the Joule heat evolution
increases together with the current. Together with other characteristic quantities $(I_{ct}, I^*, I_{b.c.}, I_m, \alpha$, etc.) the minimum normal-
zone propagation current I_p is one of the most important characteristics of the combined conductor.

Thus a study of the state of equilibrium in a combined conductor subject to isothermal conditions should certainly be supplemented by a study of the equilibrium conditions (i.e., the conditions
of existence of the normal zone) under circumstances in which this
zone occupies only part of the length of the superconductor. If a
normal zone extends over a finite length of the combined conductor,
the temperature gradient along the conductor will differ from zero
(Fig. 5-1). On the section of the conductor occupied by the normal
zone (in which Joule heat is evolved in the presence of a current)
the temperature is higher than on the section in the superconducting state. On the first section $T > T_{c0}(B, 0)$ and on the second
$T < T_c(B, I)$; between these is a region in which the superconductor is in the resistive state.

5.2. Combined Conductor with a

Longitudinal Temperature Gradient

The problem of the propagation of a normal zone along an infinitely long combined conductor was first considered theoretically by B. N. Samoilov, M. G. Kremlev, E. Yu. Klimenko, and V. E.
Keilin [59]. In this investigation a combined conductor immersed
in a helium tank was considered; the external magnetic field was
assumed constant and homogeneous. Using values of the specific
heat c and thermal conductivity λ averaged over the whole conductor, the following equation may be set down for the propagation of heat along the conductor in the presence of heat transfer

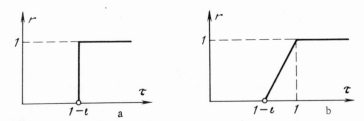

Fig. 5-2. Dependence of the dimensionless effective resistance of the combined conductor on the dimensionless temperature. a) For the first model; b) for the second model.

from the surface into the helium tank and internal Joule heat evolution:

$$\frac{\partial T}{\partial t} = a\,\frac{\partial^2 T}{\partial x^2} - \frac{hP\,(T-T_t)}{c\gamma A_\Sigma} + \frac{I^2 R}{c\gamma A_\Sigma},\qquad(5\text{-}1)$$

where x is the coordinate directed along the conductor, $a = \lambda/c\gamma$ is the thermal diffusivity, γ is the density of the material composing the conductor, R is the effective resistance of unit length, and A_Σ is the cross-sectional area of the conductor.

Equation (5-1) may be reduced to dimensionless form

$$\frac{\partial \tau}{\partial \theta} = \frac{\partial^2 \tau}{\partial X^2} - \tau + \tilde{a}\tilde{r}\iota^2,\qquad(5\text{-}2)$$

where $\theta = thP/c\gamma A_\Sigma$ is the dimensionless time, $X = x(hP/\lambda A_\Sigma)^{1/2}$ is the dimensionless coordinate, and $\tilde{r} = RA_\Sigma/\rho$ is the dimensionless effective resistance.

Two forms of the r(T) relationship (Fig. 5-2) were studied by Keilin et al. [59]. In the first case the effective resistance $\tilde{r}$ changes abruptly from 0 to 1 when $\tau^* = 1 - \iota$ (Fig. 5-2a). For this form of temperature dependence of $\tilde{r}$ it is easy to convert (5-2) into an expression for the velocity of the normal zone along the conductor:

$$c = \frac{\alpha \iota^2 + 2\iota - 2}{\sqrt{(1-\iota)\,(\alpha\iota^2 + \iota - 1)}}.\qquad(5\text{-}3)$$

An analysis of Eq. (5-3) enables us to distinguish two characteristic values of the current

$$\left.\begin{array}{l} \iota^*{}_1 = \sqrt{1/4\alpha^2 + 1/\alpha} - 1/2\alpha; \\ \iota^*{}_2 = \sqrt{1/\alpha^2 + 2/\alpha} - 1/\alpha. \end{array}\right\} \qquad (5\text{-}4)$$

For $\iota = \iota_1^*$ the rate of contraction of the normal zone tends to infinity. For $\iota = \iota_2^*$ the rate of propagation of the zone is zero; thus ι_2^* plays the part of the minimum normal-zone propagation current in the present model. It is easy to show that for a sudden change in r it is in principle impossible to obtain a completely stabilized conductor, i.e., a conductor in which the propagation of the normal zone is prevented for $I < I_c$ (see Section 5.1). It follows in fact from Eqs. (5-4) that even when $\alpha \rightarrow 0$ (very intensive cooling) the values of the currents ι_1^* and ι_2^* always remain smaller than unity.

These particular characteristics of the model under consideration arise from the fact that the model fails to allow for the division of the current between the superconductor and the normal metal.

In the second model studied by Keilin et al. [59] the temperature dependence of the resistance had the form illustrated in Fig. 5-2b. We see from the figure that for $\tau \leq \tau^*$ the electrical resistance of the combined conductor is equal to zero and the total current $\iota = I/I_c$ flows solely through the superconductor. For $\tau = \tau^* = 1 - \iota$ a resistive impedance appears in the superconductor, and for $\tau > \tau^*$ the current is divided between the superconductor and the substrate. Finally for $\tau = 1$ ($T = T_c$) the superconductor passes into the normal state and nearly all the current is transferred to the substrate, the value of $\widetilde{r}$ for $\tau \geq 1$ being equal to unity. As already mentioned, for dimensionless temperatures between τ^* and 1 the $\widetilde{r}(\tau)$ relationship is regarded as linear.

The relationship between the effective resistance and τ may be written in the following way:

$$\widetilde{r} = \begin{cases} 0 & \text{for } \tau < 1 - \iota \\ \dfrac{\tau + \iota - 1}{\iota} & \text{for } 1 - \iota < \tau < 1 \\ 1 & \text{for } \tau > 1. \end{cases} \qquad (5\text{-}5)$$

Allowing for (5-5) we may use (5-2) to find the rate of propagation of the normal phase along the conductor; however, an exact solution of this equation is more complicated. Relationships

analogous to Eqs. (5-4) were derived in Ref. 59. These determine the current ι_2 in the combined conductor for which the rate
of propagation of the normal zone through the superconductor is
equal to zero as a function of the Stekly stability criterion:

$$\iota_2 = \sqrt{\frac{1}{4\alpha^2} + \frac{2}{\alpha}} - \frac{1}{2\alpha}; \qquad (5-6)$$

furthermore the current ι_1, which in the present model coincides
with the current ι_m, introduced earlier is equal to

$$\iota_1 = \sqrt{\frac{1}{\alpha}} = \iota_m. \qquad (5-7)$$

It follows from Eq. (5-6) that $\iota_2 = 1$ (i.e., $I_2 = I_{ct}$) for $\alpha = 1$.
Thus for this particular model the normal zone cannot in principle
exist right up to the critical current in a completely stabilized
conductor in the absence of external sources of heat evolution. The
quantity ι_2 is the minimum normal-zone propagation current for
the model in question.

In the general case, in order to make a more complete analysis of the behavior of the normal zone in a combined conductor
with a longitudinal temperature gradient, it would be extremely
convenient to "stabilize" the normal zone over the entire range of
working currents and consider only the stationary (steady-state)
solutions of Eq. (5-1). This may be achieved by using an additional heat source of specified intensity covering a short part of the
length of the conductor.

By varying the power of the heater we may thus create a
stable normal zone in the conductor even in the current range
$\iota < \iota_m$, in which under ordinary circumstances this zone cannot
in principle exist (see Section 4.1). This presentation of the problem is of great practical significance, since in real superconducting coils there are usually a number of potential sources of additional heat evolution (for example, breaks in the superconducting
strands of the conductor, nonuniformities in the properties of the
superconductor along its length, junction points, and so on). These
sources may in principle promote the development of a normal
zone. On the other hand, in a number of cases microheaters are
specially placed in the coils of superconducting devices as a peculiar form of diagnostic instrument, providing information regard-

ing the reserve of stability of the coil, the critical current in various sections, and so on.

In the theoretical model a microheater of power W_h may be allowed for in two completely equivalent ways.

1. Considering that the length of the section of combined conductor in which the microheater resides is negligibly small so that almost all its emitted heat only travels along the conductor (symmetrically in both directions), we may write the heat-balance equation in the form

$$W_h = -2\lambda A_\Sigma \left(\frac{\partial T}{\partial x}\right)_{x=0},$$ (5-8)

where x = 0 is the site of the microheater. This equation is essentially a boundary condition for Eq. (5-2); it determines the temperature gradient at a particular point of the combined conductor.

2. We may add a term of form $W_h \delta(x)$ to the right-hand side of Eq. (5-1), where δ is a Dirac function defined as follows:

$$\delta(x) = \begin{cases} 0 & \text{for } x \neq 0 \\ \infty & \text{for } x = 0, \end{cases}$$ (5-9)

where

$$\int_{-\infty}^{\infty} \delta(x)\, dx = 1.$$

By "matching" the stationary solutions of Eq. (5-2) at the boundaries of the corresponding temperature ranges [see (5-5) and Fig. 5-2] it is quite easy to obtain the length l of the section of combined conductor which has passed into the normal state as a function of the current I in the conductor for various powers of the microheater. An example of such a calculation is presented in Section 5.3.

Figure 5-3a shows the $L-\iota$ characteristic for the case of a constant heat-transfer coefficient h. The following notation is used in the figure: $L = l/l_0$ is the dimensionless length of the section of combined conductor in the normal and/or resistive state $l_0 = (\lambda A_\Sigma/hP)^{1/2}$; $f = W_h/W_{h0} = -\partial \tau/\partial X$ is the dimensionless heater power; and W_{h0} is the power of the microheater for which

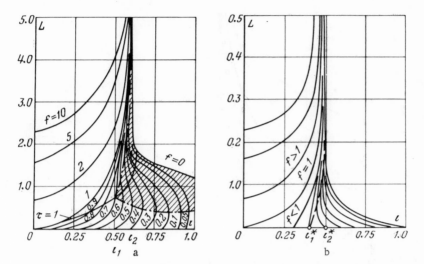

Fig. 5-3. The $L-\iota$ characteristics of the combined conductor (a) for a linear $r(\tau)$ relationship and (b) for an abrupt $r(\tau)$ relationship.

the temperature at its point of installation is equal to $T_c(B, 0)$ in the absence of a current.

The stability parameter α was taken as equal to 4.

It is easy to recognize the two characteristic currents ι_1 and ι_2 in Fig. 5-3a. For $\iota < \iota_1$ the derivative $(\partial L/\partial \iota)_f$ is always positive while for $\iota > \iota_1$ in certain regions the curves may have a negative slope. As $\iota \to \iota_2$ the length of the normal zone increases without limit.

As already indicated [59], $\iota_1 = 1/\sqrt{\alpha}$, i.e., the quantity ι_1 coincides with the minimum current corresponding to the existence of the normal phase ι_m. It is easy to see that under steady-state conditions an infinitely long normal zone (in one direction) corresponds to precisely the minimum normal-phase propagation current; hence

$$\iota_2 = \iota_p = \sqrt{\frac{1}{4\alpha^2} + \frac{2}{\alpha} - \frac{1}{2\alpha}}.$$

For conductors in which $\alpha < 1$ the value of ι_p exceeds 1, and for all currents up to $\iota = 1$ regions in which $(\partial L/\partial \iota)_f < 0$ are completely absent. This may readily be seen from Fig. 5-4, which shows the $L(\iota)$ relationship for the case $\alpha < 1$.

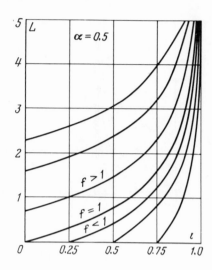

Fig. 5-4. The L(ι) relationship for the
case $\alpha = 0.5$.

Special analysis shows that states of the combined conductor
in which $(\partial L/\partial \iota)_f < 0$ are unstable. In Fig. 5-3a the regions of
unstable states are shaded. Of course for $f = 0$ any states except
$\iota = \iota_p$ are unstable; the normal or resistive zone either extends
or contracts in length.

It is interesting to note that for $\iota = 1$ and $f = 0$ there is a section of the combined conductor in the resistive state. For comparison Fig. 5-3b shows the L(ι) relationship relating to $\alpha = 4$,
constructed for the case in which the effective resistance $\tilde{r}(\tau)$
changes abruptly (Fig. 5-2a). We see that in contrast to the previous diagram the value of L tends to zero as $\iota \to 1$ for $f = 0$. As
in the case of a linear dependence of the effective resistance $\tilde{r}(\tau)$,
this diagram exhibits characteristic currents ι_1^* and ι_2^*.

The characteristics presented in Fig. 5-3 are very well defined and enable us to analyze the behavior of the normal zone in
a simple and convenient manner over the whole current range $0 - I_c$.

At the same time it is important to note that a comparison of
the theoretical diagram in Fig. 5-3a with experiment, in which we
are usually concerned with the volt−ampere characteristics,* is
rather difficult, since when the current is divided between the
superconductor and the substrate the voltage on the sample is not

---

*It is experimentally very difficult to secure a direct change in the length of the nor-
mal (resistive) section for different currents in a combined conductor.

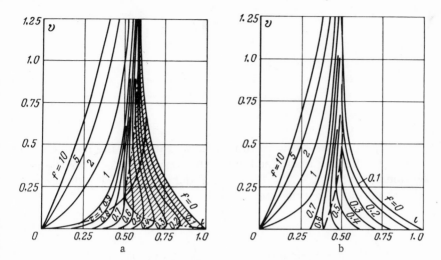

Fig. 5-5. Volt—ampere characteristics in the presence of a temperature gradient
(α = 4).

proportional to the length of the resistive section. In this connection we also constructed the theoretical dimensionless voltage
drop in the normal zone as a function of ι for α = 4 and f = const
(Fig. 5-5a). It is interesting to note that when f = 0 the voltage
drop tends to zero for $I = I_c$, although the length of the resistive
section remains finite (see Fig. 5-3b).

For the case in which the effective resistance $\tilde{r}(\tau)$ changes
abruptly the volt—ampere characteristic is obtained on simply
multiplying L by ι for the corresponding values of f (Fig. 5-5b).

5.3. Effect of the Boiling Crisis

on the Equilibrium Conditions

The calculation described in the preceding section was carried out on the assumption that the heat-transfer coefficient h was
independent of temperature. However, as already indicated in
Chapter 4 for the case of equilibrium under isothermal conditions,
allowance for the h(T) relationship, and most of all the sharp
change in the heat-transfer coefficient resulting from the boiling
crisis, is of very great practical importance.

Complete allowance for the exact temperature dependence of
the heat-transfer coefficient is difficult and hardly appropriate;

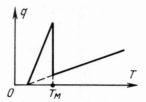

Fig. 5-6. Linearized approximation for the q(T) relationship.

it is perfectly reasonable to limit consideration to the boiling crisis by using the linearized approximation (Fig. 5-6) for the q(T) relationship when $T < T_M$ (h = const) and when $T > T_M$ (h' = const).
Let us introduce the notation

$$h/h' = \gamma^2. \tag{5-10}$$

From Eq. (5-2) for steady-state conditions $(\partial \tau/\partial \theta = 0)$ we have for $\tau < \tau_M$

$$\frac{\partial^2 \tau}{\partial X^2} - \gamma^2 \tau + a\iota^2 r(\tau) = 0; \tag{5-11}$$

for $\tau > \tau_M$

$$\frac{\partial^2 \tau}{\partial X^2} - \tau + a\iota^2 r(\tau) = 0. \tag{5-12}$$

Here

$$\tau_M = \frac{T_M - T_t}{T_{c0} - T_t}. \tag{5-13}$$

This quantity depends on B since T_c is a function of B. As before we assume that the $\tilde{r}(\tau)$ relationship is determined by Eq. (5-5).

We shall solve Eqs. (5-11) and (5-12) by a method analogous to that used in Ref. 59, i.e., by finding solutions for individual intervals of τ inside which $\tilde{r}(\tau)$ is either constant or varies linearly. The resultant solutions are "matched"("sewn together") at the boundaries of the intervals, as a result of which l/l_0 and $UA/\rho\, I_{ct}$ are determined as functions of ι for f = const.

Depending on the power of the microheater and the value of the current the following situations may arise:

1. If the power of the microheater is fairly low, the entire superconductor in the combined conductor will be in the superconducting state;

2. After the power of the microheater has exceeded a certain value the section of superconductor close to the site of the latter will be in the resistive state, and outside this region in the superconducting state;

3. On further raising the power of the microheater a normal zone will arise close to its position; on both sides there will be a resistive zone, and beyond these the superconductor will be in the superconducting state.

It is very obvious that for different values of the current ι there may be different relationships between the value of τ_M (not depending on the current) and the lower temperature boundary of the resistive zone $\tau = 1 - \iota$, and also the lower temperature boundary of the normal zone $\tau = 1$, which naturally forms the upper boundary of the resistive zone.

The boiling crisis $(\tau = \tau_M)$ may take place in part of the superconducting $(\tau_M < 1 - \iota)$, resistive $(1 - \iota < \tau_M < 1)$, or normal zone $(\tau_M > 1)$. In the latter case the relationship between τ_M and $\tau = 1$ is naturally independent of current.

The temperature distribution along the combined conductor for these cases is illustrated schematically in Fig. 5-7. The results of the calculation based on Eqs. (5-11) and (5-12) are different according to the values of α, γ^2, and τ_M.

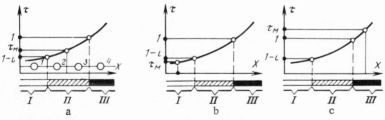

Fig. 5-7. Temperature distribution along the length of the combined conductor for τ = const. (a) $1 - \iota < \tau_M < 1$; (b) $\tau_M < 1 - \iota$; (c) $\tau_M > 1$. I) Superconducting section; II) resistive section; III) normal section.

In practice we almost always have $\tau_M < 1$, which corresponds to $T_{c0} > 5$ K; we need therefore only consider two of the cases mentioned, those relating to $1 - \iota < \tau_M < 1$ and $\tau_M < 1 - \iota$.

1. Case $1 - \iota < \tau_M < 1$, i.e, the boiling crisis occurs in the resistive zone (Fig. 5-7a). The solutions of Eqs. (5-10) and (5-11) will be different for four sections of the combined conductor:

Section 1: $\tau \le 1 - \iota < \tau_M$ (superconducting state, bubble-type boiling). In this section $\tilde{r}(\tau) = 0$ and hence Eq. (5-11) may be written

$$\frac{\partial^2 \tau}{\partial X^2} - \gamma^2 \tau = 0, \tag{5-14}$$

whence

$$\tau = \tau_0 e^{\gamma X}. \tag{5-15}$$

Let us take as the initial point for evaluating the coordinate the initial point of the resistive zone:

$$\tau = \tau_0 = 1 - \iota \quad \text{for} \quad X = 0. \tag{5-16}$$

Allowing for this condition Eq. (5-15) may written in the form

$$\tau = (1 - \iota) e^{\gamma X}. \tag{5-17}$$

The temperature gradient at the point $X = 0$ is determined by

$$\left(\frac{\partial \tau}{\partial X}\right)_{\tau=1-\iota} = \gamma(1 - \iota). \tag{5-18}$$

Section 2: $1 - \iota \le \tau \le \tau_M < 1$ (resistive state, bubble-type boiling).

It follows from (5-13) and Fig. 5-2 that for $1 - l < \tau < 1$

$$\tilde{r}(\tau) = \frac{\tau + \iota - 1}{\iota}. \tag{5-19}$$

Hence for the section under consideration Eq. (5-11) takes the form

$$\frac{\partial^2 \tau}{\partial X^2} - \gamma^2 \tau + \alpha \iota (\tau + \iota - 1) = 0. \tag{5-20}$$

The general solution of this equation may be written as follows:

$$\tau = A \cosh(\sqrt{\gamma^2 - \alpha\iota}\,X) + B \sinh(\sqrt{\gamma^2 - \alpha\iota}\,X) + \frac{\alpha\iota\,(\iota - 1)}{\gamma^2 - \alpha\iota}. \qquad (5\text{-}21)$$

The constants A and B are determined by means of the boundary conditions at the initial point of the section derived earlier for this point (as the end of section 1). In this way "matching" of the solutions is ensured for sections 1 and 2. We obtain

$$A = 1 - \iota - \frac{\alpha\iota\,(\iota - 1)}{\gamma^2 - \alpha\iota}; \quad B = \frac{\gamma\,(1 - \iota)}{\sqrt{\gamma^2 - \alpha\iota}}. \qquad (5\text{-}22)$$

Let us use l to denote the length of section 2, i.e., the length of a segment of the resistive zone from its origin to that point at which the temperature reaches τ_M and the boiling crisis accordingly takes place:

$$L_2 = X_{\tau=\tau_M}. \qquad (5\text{-}23)$$

We find the value of L_2 from Eq. (5-21) for $\tau = \tau_M$:

$$L_2 = \frac{1}{\sqrt{\gamma^2 - \alpha\iota}} \sinh^{-1} \frac{B\left[\tau_M - \frac{\alpha\iota\,(\iota - 1)}{\gamma^2 - \alpha\iota}\right] - A\sqrt{B^2 - A^2 + \tau_M - \frac{\alpha\iota\,(\iota - 1)}{\gamma^2 - \alpha\iota}}}{B^2 - A^2}. $$

$$(5\text{-}24)$$

The temperature gradient at the point $X = L_2$ is

$$\left(\frac{\partial\tau}{\partial X}\right)_{\tau=\tau_M} = A\sqrt{\gamma^2 - \alpha\iota}\,\sinh(\sqrt{\gamma^2 - \alpha\iota}\,L_2) + \gamma\,(\iota - 1)\cosh(\sqrt{\gamma^2 - \alpha\iota}\,L_2). $$

$$(5\text{-}25)$$

Section 3: $\tau_M \leq \tau \leq 1$ (resistive state, film-type boiling). For this section Eq. (5-12) is written in the form

$$\frac{\partial^2\tau}{\partial X'^2} - \tau + \alpha\iota\,(\tau + \iota - 1) = 0. \qquad (5\text{-}26)$$

The quantity X' in this equation is related to the X in Eqs. (5-17) and (5-21) in the following way:

$$X' = X - L_2. \qquad (5\text{-}27)$$

The general solution of Eq. (5-26) for the case $\alpha \iota - 1 > 0$ takes the following form:

$$\tau = C \cos\left(\sqrt{\alpha \iota - 1} X'\right) + D \sin\left(\sqrt{\alpha \iota - 1} X'\right) + \frac{\alpha \iota (\iota - 1)}{1 - \alpha \iota}. \qquad (5\text{-}28)$$

When $\alpha \iota - 1 < 0$ the trigonometric functions in this equation are replaced by hyperbolic functions.

The constants C and D are determined by means of the boundary conditions at the initial point of section 3, i.e., at the end of section 2 (X = L$_2$). The first of these conditions lies in the fact that at X = L$_2$ (i.e., at X' = 0):

$$\tau = \tau_M, \qquad (5\text{-}29)$$

the second condition being given by (5-25). Thus matching of the solutions for sections 2 and 3 is ensured. We obtain

$$C = \tau_M - \frac{\alpha \iota (\iota - 1)}{1 - \alpha \iota}; \quad D = \frac{1}{\sqrt{\alpha \iota - 1}} \left(\frac{\partial \tau}{\partial X}\right)_{\tau = \tau_M}. \qquad (5\text{-}30)$$

Let us use L$_3$ to denote the length of the section of combined conductor in which the superconductor has a resistive impedance, i.e., the length of the section corresponding to the temperature range $\tau (1 - \iota, 1)$.

Clearly we have

$$\tau(L_3) = 1. \qquad (5\text{-}31)$$

It follows from Eq. (5-28) that

$$\left(\frac{\partial \tau}{\partial X}\right)_{\tau=1} = -A\sqrt{4\iota - 1} \sin\left[\sqrt{4\iota - 1}\,(L_3 - 1)\right] +$$

$$+ B\sqrt{4\iota - 1} \cos\left[\sqrt{4\iota - 1}\,(L_3 - 1)\right]. \qquad (5\text{-}32)$$

The quantity (L$_3$ − L$_2$) is determined from Eq. (5-28) at $\tau = 1$, while L$_3$ is obtained by adding the result to the value of L$_2$ found earlier.

For any particular power of the microheater there exists a (unique) value of the current ι for which part of the combined conductor is occupied by a resistive zone at the boundaries of which the temperatures are $\tau = 1 - \iota$ and $\tau = 1$. The second of these values of τ corresponds to the point at which the microheater is situated. In accordance with the foregoing discussion the length of the resistive zone is L.

Since the dimensionless power of the microheater f is related to the dimensionless temperature gradient at the site of the microheater by

$$f = -\frac{\partial \tau}{\partial X}, \tag{5-33}$$

for a fixed value of f the quantity ι is determined by means of Eq. (5-32) from the known value of $(L_3 - L_2)$.

The length of the section of combined conductor occupied by the resistive zone for smaller currents and a specified value of f is given by Eq. (5-32), in which the value of L_3 is replaced by the current value of $X(\iota) < L_3$, and may also be obtained from Eq. (5-24).

In this way for lines of f = const we may determine the current dependence of the section of combined conductor having a resistance of

$$X(\iota) = L_2 + (X - L_2), \tag{5-34}$$

for temperatures $\tau \leq 1$.

Section 4: $\tau \geq 1$ (normal state, film-type boiling).

Since when $\tau \geq 1, \tilde{r}(\tau) = 1$, Eq. (5-11) may be written for this section as follows:

$$\frac{\partial^2 \tau}{\partial X''^2} - \tau + \alpha \iota^2 = 0. \tag{5-35}$$

The quantity X'' is related to X in the following way:

$$X'' = X - L_3. \tag{5-36}$$

The general solution of Eq. (5-35) takes the form

$$\tau = M \cosh X'' + N \sinh X'' + \alpha \iota^2. \tag{5-37}$$

The constants M and N are determined by means of the boundary conditions at the initial point of section 4, i.e., the end of section 3 ($X = L_3$), by virtue of conditions (5-31) and (5-32); these equations ensure the "matching" of the solutions for sections 3 and 4. We obtain

$$M = 1 - \alpha \iota^2; \quad N = \left(\frac{\partial \tau}{\partial X}\right)_{\tau=1}. \tag{5-38}$$

Let us use L to denote the length of the section of combined conductor in which the superconductor has a resistance, i.e., the total length of the resistive and normal zones. The temperature gradient at the site of the microheater X = L, equal in absolute magnitude to the dimensionless power of the heater, is given by

$$\left(\frac{\partial \tau}{\partial X}\right)_{X=L} = -f = (1 - \alpha \iota^2) \sinh(L - L_3) + \left(\frac{\partial \tau}{\partial X}\right)_{\tau=1} \cosh (L - L_3), \tag{5-39}$$

where $(\partial \tau / \partial X)_{\tau=1}$ is determined by Eq. (5-32).

Equation (5-39) expresses the dependence of the quantity $(L - L_3)$ on the current for a fixed value of f. Thus for different values of ι and f = const, L is determined as the sum of L_2, $(L_3 - L_2)$, and $(L - L_3)$ respectively, determined by means of Eqs. (5-24), (5-28), and (5-39).

It should be emphasized that we have so far everywhere considered half of the combined conductor, lying on one side of the microheater. Since the temperature distribution along the length of the conductor is symmetrical with respect to the heater, the value of L found in the way just described has to be doubled.

II.   Case $\tau_M < 1 - \iota$, i.e., the boiling crisis occurs in the superconducting section of the combined conductor (Fig. 5-7b).

The solution is derived in the same way as that just considered. The combined conductor is divided into four sections along its length:

$\tau \le \tau_M \le 1 - \iota$ (superconducting section, bubble-type boiling);
$\tau_M \le \tau \le 1 - \iota$ (superconducting section, film-type boiling);
$1 - \iota \le \tau \le \iota$ (resistive section, film-type boiling); $\tau \ge 1$ (normal section, film-type boiling).

The solutions of Eqs. (5-11) and (5-12) for each of the sections are "matched" to each other in the manner already described.

As before we take the initial point of the resistive zone ($\tau = 1 - \iota$) as origin for the coordinate X and use L_1 to denote the length of the segment of superconducting zone in which film-type boiling occurs, i.e., the length of the segment corresponding to the temperature range $\tau_M \leq \tau \leq 1 - \iota$.

For section 1 ($\tau \leq \tau_M$, i.e., $X \leq L_1$) Eq. (5-11) takes the form

$$\frac{\partial^2 \tau}{\partial X^2} - \gamma^2 \tau = 0, \tag{5-40}$$

whence

$$\tau = \tau_M e^{\gamma (X + L_1)}. \tag{5-41}$$

At the boundary of this section (at $X = -L_1$)

$$\tau = \tau_M; \tag{5-42}$$

$$\left(\frac{\partial \tau}{\partial X}\right)_{\tau = \tau_M} = \gamma \tau_M. \tag{5-43}$$

For section 2 ($\tau_M \leq \tau \leq 1 - \iota$, i.e. $-L_1 \leq X \leq 0$) Eq. (5-12) is written as follows:

$$\frac{\partial^2 \tau}{\partial X^2} - \tau = 0. \tag{5-44}$$

The first integral of this equation may be expressed in the form

$$\left(\frac{\partial \tau}{\partial X}\right)^2 - \tau^2 + E = 0. \tag{5-45}$$

The constant E is determined by means of boundary conditions (5-42) and (5-43):

$$E = (1 - \gamma^2) \tau_M^2. \tag{5-46}$$

For $X = 0$ we have

$$\tau = 1 - \iota; \tag{5-47}$$

$$\left(\frac{\partial \tau}{\partial X}\right)_{\tau = 1 - \iota} = \sqrt{(1 - \iota)^2 + (\gamma^2 - 1)\, \tau_M}. \tag{5-48}$$

The determination of the solutions for sections 3 and 4 and the "matching" of these are carried out in the same way for sections 3 and 4 in case I. In this way the $L(\iota)$ relationship for $f =$ const is determined with due allowance for the helium boiling crisis.

Using the known $L(\iota)$ relationship it is extremely easy to find the current dependence of $UA/\rho I_c$, the dimensionless voltage drop in the section of the combined conductor possessing a resistance. If L is the length of the section under consideration, then the voltage drop for this section is found as the sum of the values of v for the resistive section of length L_3 and the normal section of length $(L - L_3)$:

$$v = v_{res} + v_{norm} \qquad (5\text{-}49)$$

The voltage drop in unit length of the combined conductor is determined by the equation

$$v = \tilde{r}\iota. \qquad (5\text{-}50)$$

Since in the resistive part of the section in question $\tilde{r}(\tau)$ is determined by Eq. (5-19), for this part

$$v_{res} = \tau + \iota - 1. \qquad (5\text{-}51)$$

Thus

$$v_{res} = \int_0^L \frac{UA}{\rho I_{ct}} \, dX, \qquad (5\text{-}52)$$

or

$$v_{res} = \int_0^L \tau \, dX + (\iota - 1) L_3. \qquad (5\text{-}53)$$

The voltage drop for the normal part of the section is

$$v_{norm} = \int_{L_3}^L \frac{UA}{\rho I_{ct}} \, dX. \qquad (5\text{-}54)$$

Since for the normal zone $\tilde{r} = 1$, and hence, in accordance with
(5-50), $v = \iota$, we obtain

$$v_{norm} = \iota\,(L - L_3). \qquad (5\text{-}55)$$

Thus

$$v = \int_0^{L_3} \tau\,(X)\,dX + \iota L - L_3. \qquad (5\text{-}56)$$

The results of our calculations of the characteristics of the
combined conductor in the presence of a longitudinal temperature
gradient (allowing for the boiling crisis) are presented in Fig. 5-8.
These data refer to a conductor in which $\alpha' = 4$, $\gamma^2 = 16$ (and hence
$\alpha = 0.25$), and $\tau_M = 0.25$. As before (see Fig. 5-3), Fig. 5-8 contains
lines constituting the geometrical locus of states in which the temperature τ at the heater site is equal to unity (for a specified
value of f). The regions below this line correspond to states for
which the superconducting zones in the combined conductor are
only accompanied by a resistive zone, the normal zone being absent.
Above this line is a region of states in which a normal as well as
a resistive and superconducting zone occurs.

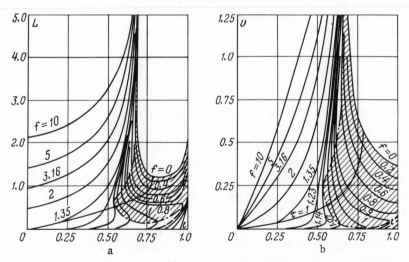

Fig. 5-8. Characteristics of a combined conductor in the presence of a longitudinal
temperature gradient, allowing for the boiling crisis.

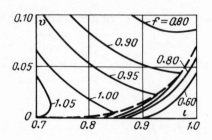

Fig. 5-9. Volt—ampere characteristics of a combined conductor close to $\iota = 1$ for small values of U.

A comparison between the results presented in Figs. 5-3 and 5-8 shows that making an allowance for the boiling crisis leads to a considerable (quantitative and qualitative) change in the calculated characteristics of the combined conductor. The most noticeable differences occur in the region $\iota < \iota_p$, especially close to $\iota = 1$.

Allowance for the boiling crisis has the effect that for $\iota = 1$ and a number of values of $f \geq 0$ the quantity U is positive (Fig. 5-8). This result, however, is in complete conformity with the volt—ampere characteristic of the combined conductor for the case $\alpha' < 1$. This correspondence may be understood if we remember that in the range of currents close to $\iota = 1$ the dimensionless temperature of the conductor τ tends to zero. In this temperature range when $\tau < \tau_M$ Eq. (5-11) is valid, and the part of the parameter α' is played by the quantity $\alpha = \alpha'/\gamma^2$.

The change in the lines f = const close to $\iota = 1$ for small values of U is illustrated in Fig. 5-9, in which the lower right-hand part of Fig. 5-8b is shown on a large scale. In this figure for a specified value of f the broken line represents the geometrical locus of states in which the temperature of the combined conductor at the site of the microheater is equal to τ_M, i.e., the temperature at the onset of the boiling crisis. The region below this line corresponds to states in which the boiling of the helium on the surface of the combined conductor is solely of the bubble type. Above this line is a region of states in which film-type boiling occurs on part of the surface of the conductor. Close to the points of transition through this line the f = const curves bend sharply to the left.

In analyzing the data presented in Figs. 5-3a, 5-5b, and 5-8, we should direct attention to the character of the f = 0 line, i.e., the line corresponding to the equilibrium conditions in the absence of an additional point source of heat evolution. This line

corresponds to the existence of a normal zone under the real practical conditions of a combined conductor. We see from Figs. 5-3
and 5-8 that this line only exists in the current range between
$\iota = \iota_p$ and $\iota = 1$. For $\iota < \iota_p$ the normal phase cannot exist under
these conditions if it is not supported by an additional microheater.
The states corresponding to $f = 0$ in the current range $\iota = \iota_p$ to
$\iota = 1$ are unstable: In the presence of a longitudinal temperature
gradient the normal zone developing either extends without limit
along the conductor or else "collapses."

As already noted, for $\iota = 1$ and $f = 0$ the voltage drop along
the entire length of the conductor in the resistive state is finite;
for the conductor represented by Fig. 5-8b we have $v_{i=1,\,f=0} =$
0.363. Neglect of the boiling crisis leads to a completely different result; we see from Fig. 5-5 that $v_{i=1,\,f=0} \approx 0$.

It is also clear from Fig. 5-3a that for $\iota = 1$ and $f = 0$ the
length of the resistive section tends to a finite value in the absence
of a boiling crisis, while the vanishingly small voltage drop in
this section is explained by the extremely low resistivity of the
superconductor in the resistive state (for $\iota = 1$).

A more general approach to the study of thermal equilibrium and the propagation of the normal zone in combined conductors proposed by Maddock et al. [56] is
also possible. The essence of this method is as follows.

Let us consider an infinitely long conductor in which heat evolution W occurs
and heat outflow q takes place from the surface, W and q being unique functions of
the surface temperature T. Let us assume that these functions have several points of
intersection (equilibrium points), in which the heat evolution is naturally equal to the
heat outflow (Fig. 5-10a):

$$W(T) = q(T). \tag{5-57}$$

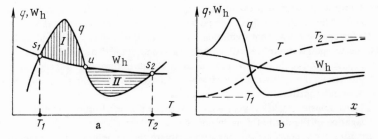

Fig. 5-10. Relationship between the heat evolution W, the heat outflow q,
the surface temperature of the conductor T (a), and the x coordinate (b).

Using our earlier methods developed for examining the stability of the states (Section 4.4) it is easy to show that the equilibrium states at temperatures T_1 and T_2 (points s_1 and s_2) are stable in the isothermal case, since they satisfy condition (4-93). Conversely the state at the intersection u is unstable, since at this point condition (4-94) is satisfied.

Figure 5-10b shows the same functions W(T) and q(T) as functions of coordinate x measured along the conductor.

The thermal-balance equation for an element of the conductor may be written in the following form:

$$\frac{A}{P} \frac{d}{dx} \left(\lambda \frac{dT}{dx} \right) = q(T) - W(T),$$ (5-58)

where x is the coordinate along the conductor, A is the cross section, P is the cooled perimeter, and λ is the thermal conductivity of the conductor material.

We require to find a solution of Eq. (5-58) such that for the case in which one end of the conductor is at temperature T_1 and the other at T_2 there is a stable temperature profile in the temperature range T_1 to T_2.

Let us introduce the new variable $S = \lambda(dT/dx)$. Equation (5-58) then easily transforms to the following:

$$\frac{A}{P} \int S \, dS = \int [q(T) - W(T)] \lambda(T) \, dT.$$ (5-59)

It is quite obvious that for a fairly long conductor with boundary temperatures T_1 and T_2 the integration limits for S on the left-hand side of Eq. (5-59) will vanish. The equation will thus take the form

$$\int_{T_1}^{T_2} [q(T) - W(T)] \lambda(T) \, dT = 0.$$ (5-60)

If the thermal conductivity λ does not depend on temperature,

$$\int_{T_1}^{T_2} [q(T) - W(T)] \, dT = 0.$$ (5-61)

Equation (5-61) means that in order to ensure a stable temperature profile along the conductor in question it is essential for the areas of regions I and II in Fig. 5-10a to be equal. This conclusion has become known as the "theorem of equal areas."

If, for example, the heat evolution W(T) is smaller than that implied by Eq. (5-61), the temperature profile will change shape and "move" along the conductor until the conductor reaches a new state of thermal equilibrium below the point s_1. If W(T) increases, the temperature profile will "move" to the right, reaching a new state of thermal equilibrium above the point s_2.

This conclusion makes it very easy to analyze the equilibrium conditions and the propagation of the normal zone in combined conductors.

By using these concepts we may also estimate the velocity of the temperature profile, or (amounting to the same thing) the rate of propagation of the normal zone along the conductor. For the case in which the origin of coordinates moves at a velocity $\mathfrak{v}$ equal to the velocity of the temperature profile along the conductor, the heat-balance equation (5-58) may be written

$$S \frac{dS}{dT} = [q(T) - W(T)] \frac{P\lambda}{A} + \mathfrak{v}cS, \qquad (5\text{-}62)$$

where c is the specific heat of the conductor material.

Since the temperature gradient at the ends of the conductors with $T = T_1$ and $T = T_2$ is equal to zero, after integration Eq. (5-62) is transformed as follows:

$$\int_{T_1}^{T_2} [q(T) - W(T)] \frac{P\lambda}{A} dT + \mathfrak{v} \int_{T_1}^{T_2} cS \, dT = 0. \qquad (5\text{-}63)$$

Thus the velocity of the temperature profile is given by

$$\mathfrak{v} = - \frac{P \int_{T_1}^{T_2} [q(T) - W(T)] \lambda \, dT}{A \int_{T_1}^{T_2} cS \, dT}. \qquad (5\text{-}64)$$

Using Eq. (5-64) and the relationships q(T), W(T), and λ(T) already known for each particular case, by successive approximations we may find the numerical values of the velocity of the temperature profile along the conductor.

The method based on the use of the "theorem of equal areas" is especially convenient; it enables us to make a very simple analysis of the equilibrium states of combined conductors in various situations in a particularly explicit form.

By way of example, Fig. 5-11 presents a diagram which easily enables us to find the states of thermal equilibrium and hence the characteristic currents for a combined conductor with a stabilization parameter $\alpha' > 1$ in the presence of the boiling crisis in the liquid helium. The broken line characterizes the heat outflow q from the surface of the conductor into the tank containing the liquid helium. The continuous lines indicate the heat evolution W for current values of $\iota_1 < \iota_2 < \iota_3 < \iota_4$.

Since the power of the heat evolution in the conductor is equal to $\alpha \iota^2 r$, it is quite obvious that the heat-evolution curves for each current value have the same form as the temperature dependence of the dimensionless resistance of the conductor r (see Fig. 5-2b), defined by Eq. (5-5).

We see from Fig. 5-11 that for a current ι the conductor is completely stabilized, since there are no states of thermal equilibrium between the conductor and the ambient (the W and q curves nowhere intersect, the W curve being situated below the heat-transfer curve q). This situation persists up to a current of ι_2, at which time the curves in question have a common point at a temperature $\tau = 1$. This current value may be defined as the minimum current for the existence of a normal zone (Section 4.1). On further increasing the current to $\iota = \iota_3$ a new stable state of equilibrium is created in which the area I is equal to the area II. In other words, at

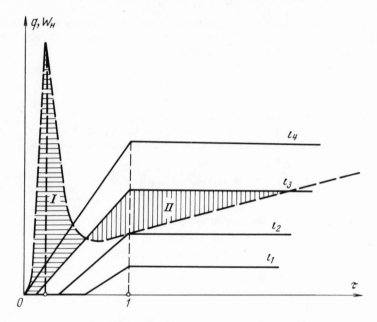

Fig. 5-11. Curves for determining the states of thermal equilibrium for $\alpha' > 1$
in the presence of the boiling crisis.

$\iota = \iota_3$ the principal condition of the "theorem of equal areas" is satisfied. Clearly the current ι_3 in this case coincides with the minimum normal-zone propagation current.

The current ι_4 is the critical current for the particular conductor at the temperature of the helium tank $\tau = 0$. By considering the heat evolution and heat-transfer curves W and q for this case it is clear that at $\tau = 0$ the conductor is completely stabilized with respect to the bubble type of boiling, since

$$(dq/d\tau)_{\tau=0} > (\partial W/\partial \tau)_{\tau=0}.$$

However, if for any reason the temperature on the surface of the conductor exceeds the temperature of the boiling crisis, the conductor becomes unstabilized with respect to the film type of boiling, and the normal zone extends in an uncontrollable manner along the conductor (since the area II then exceeds the area I).

The "theorem of equal areas" may easily be generalized to the case in which the thermal conductivity of the conductor is proportional to the absolute temperature. We readily see from Eq. (5-60) that in this case it is sufficient to represent the heat outflow (transfer) and heat evolution in the conductor as functions of the square of the temperature.

The method here considered may also be used for studying the influence of contact resistance at the superconductor−substrate boundary, the dimensions of the superconducting strands, and other factors on the state of thermal equilibrium of the combined conductors.

5.4. Effect of Electrical Contact Resistance

In Chapter 4 we considered the influence of contact (or transient) thermal resistance due to the imperfect contact between the adjacent metals, oxide films, contaminants, etc., on the thermal stability of the conductor. It is clear that all these causes should lead to the appearance of a finite electrical resistance at the superconductor—substrate boundary — also called the contact (transient) resistance. There are also certain combined conductors in which, for particular technological reasons, the superconductor is connected to the stabilizing material via an alloy with a relatively low electrical conductivity. Thus in twisted cables the individual copper and superconducting filaments are often bound together with ordinary tin—lead solders. Sometimes a superconducting wire is made encased in protective shells of other alloys; in these it undergoes all the necessary technological operations, the shell or sheath remaining intact while preparing combined conductors from this wire.

In all such cases the question naturally arises as to how far the existence of the contact resistance may influence the thermal stability of the conductor. This question could not be studied earlier in Chapter 4, when we were only considering a uniform temperature distribution along the conductor. The point is that, while the conditions along the length of the wire remain constant, there is no exchange of current between the superconducting and normal components. Such a cross-flow of currents is only encountered when the conditions vary along the conductor, for example, in the presence of temperature drops. Additional losses then occur in the transient electrical resistance, and these can only worsen the stability of the conductor. However, present considerations readily show that the worsening of stability due to these losses is never very serious.

The minimum equilibrium current ι_m, in fact, cannot depend on the contact resistance at all, and thus all the stability criteria based on the calculation of this current should remain valid. The propagation current ι_p should of course decrease; it is nevertheless clear from Section 5.3 [56] that ι_p will always exceed ι_m.

Let us estimate the effect of the contact resistance on the value of ι_p quantitatively. First of all it is necessary to introduce the corresponding numerical criterion characterizing the contact resistance. Let us assume that a thin uniform layer of material of thickness δ with a resistivity ρ_t is arranged around the whole perimeter of the superconductor P. A surface resistance of $R_t = \rho_t \delta / P$ will then apply to each unit length of the conductor. The use of this parameter is not always convenient, since the geometrical characteristics and resistivity of the transitional layer are often unknown. A more convenient quantity, easy to determine experimentally, is the so-called "transition length"

$$l_t = \sqrt{\frac{R_t}{\rho/A}}, \qquad (5\text{-}65)$$

where ρ/A is the linear resistance of the normal metal.

The meaning of the term "transition length" may be defined as follows. If a current is applied to the normal metal, it is not immediately transferred to the superconductor but only in stages, so that over a distance l_t the current in the normal metal diminishes by a factor of e times. In order to determine l_t it is in this case sufficient to measure the potential distribution along the conductor by means of several auxiliary contacts.

The length l_t still cannot act as a quantitative characteristic for the influence of the resistance R_t on the stability. A measure of this influence is the ratio of l_t to the characteristic length $l_0 = (\lambda A / hP)^{1/2}$ (Section 5.2), which gives the longitudinal scale in the present problem:

$$\Delta = l_t \bigg/ \sqrt{\frac{\lambda A}{hP}} = \sqrt{\frac{R_t hP}{\lambda \rho}}. \tag{5-66}$$

If in fact the reduced ratio is small, then, since substantial changes only take place in the varying quantities (temperature, critical current, etc.) over distances of the order of $(\lambda A / hP)^{1/2}$ in the conductor, the influence of the transient resistance is slight, and conversely. If the transient resistance is associated with any metallic coatings, solders, etc., the quantity Δ will be small, and in any case will never exceed unity [54]. For small values of Δ we may solve the corresponding approximate equations, and as we should expect on the basis of the foregoing considerations the influence of Δ on the current ι_p^* is extremely weak. For reference purposes we may here give the corresponding equation [54]

$$\iota_p^* = \iota_p \left[1 - \frac{2}{3} \frac{\iota_p (1 - \iota_p)}{(4 - \iota_p)} \Delta^2 \right], \tag{5-67}$$

where ι_p is the propagation current calculated without allowing for the contact resistance. Extrapolating this formula for a value of $\Delta = 1$ we may readily convince ourselves that the propagation current diminishes by no more than 5% as a result of the action of the contact resistance. Any further reduction in the current ι_p with increasing Δ should be very slight, since ι_p cannot be smaller than ι_m.

It follows that the use of different solders and protective alloys with high electrical resistances in combined conductors encounters no special obstacles.

It should nevertheless be remembered that in the presence of an electrical resistance in the conductor there will also necessarily be a thermal contact resistance capable of seriously worsening the stability of the conductor. In order to estimate the value of this resistance we set up the following comparison.

By way of example we once again consider a superconductor having a thin coating of material with low electrical and thermal conductivity encased in a common stabilizing matrix of cross section A. The dimensionless criterion which reflects the relative role of these two factors is in this model the ratio of the corresponding dimensionless resistances, i.e.,

$$\alpha / \Delta^2 = \frac{\rho_n l_c^2}{A (T_{c0} - T_0)} \frac{b}{P \lambda_t} \bigg/ \frac{\rho_t bhPA}{\pi d \rho_n \lambda_n K} = \alpha \frac{\lambda_n \rho_n}{\lambda_t \rho_t}. \tag{5-68}$$

Here the index n relates to the normal metal and t to the transitional layer.

If the intermediate coating consists of a pure metal, then on the right-hand side of the equation the ratio of the constants characterizing the properties of the material is close to unity. The value of α which should be associated with the film type of boiling at temperatures close to T_c is usually slightly greater than unity, so that the thermal resistance increases rather faster than the electrical resistivity. This relationship may be reversed for coatings of poorly conducting alloys, in which considerable deviations from the Wiedemann—Franz law may occur. However, even when the parameters under consideration are equal, the influence of the thermal resistance on, for example, the propagation current is of course the more considerable. Yet on considering the fact that for such coatings Δ is usually much smaller than unity it is reasonable to assume that even the thermal resistance will not exert any extreme influence in the present case.

In practice one sometimes encounters conductors with very long transition lengths l_t, implying a high transitional (transient) resistance. Such conductors include superconductors electrolytically coated with a thin layer of copper and enclosed in a common stabilizing matrix. The transitional resistance is in these cases associated with the presence of oxide films (insufficiently carefully removed before electroplating) on the surface of the superconductor. Extremely high contact resistance is also sometimes found in conductors stabilized with aluminum; this arises from deficiencies in the technology which allow oxide films to develop on the aluminum.

For uniform coatings of such types the relative thermal resistance α_l is still smaller, since the product $\lambda \rho$ may be many orders of magnitude greater than the corresponding value for metals. This can only have a very limited practical significance, since both electrical resistance and large transitional thermal resistances in practical conductors are associated with inhomogeneous surface inclusions, oxide films, contaminants, phase segregations, etc. The relative role of the electrical and thermal resistances in such cases is not entirely clear. It is nevertheless reasonable to assume that in view of the difference in the geometrical configurations involved in problems regarding the distribution of electrical currents and heat flows the role of thermal resistance (which has by far the greater effect, for example, on the propagation current) is the more significant in the present context also.

Thus conductors of this kind exhibiting substantial contact resistances should always be carefully investigated in order to establish a reliable idea of the permissible range of stable working currents.

It should also be remembered that there is a possibility (although not particularly obviously so at the present time) of an indirect influence of the electrical contact resistance on the stability of the conductor in cases in which the resistance of unit length is not a unique function of temperature (Section 5.5). The point is that Δ may enter into the condition determining the relationship between the currents in the two components of the conductor at the transition point, i.e., it may determine the position of this point, and hence ι_p also. Thus the experimental investigation just mentioned is also desirable even for conductors with relatively low values of Δ if these contain fairly thick superconducting wires, for which it is reasonable to assume that the internal thermal resistances are not very small.

5.5. Stability of a Combined Conductor for an Arbitrary Longitudinal Temperature Distribution

The analysis of Chapter 4 regarding the stability of the equilibrium states of a combined conductor is only valid when the temperature is constant along the whole conductor. The results obtained in Section 4.7 are thus not valid when the temperature varies along the conductor. However, even if at the onset a uniform temperature distribution is maintained in the section of sample under consideration, this may be disturbed as a result of fluctuations or the effects of the ends of the sample, which are usually subject to a variety of conditions. It is therefore important to investigate how to establish the condition of stability for the equilibrium states of the conductor in the case of arbitrary temperature distributions.

In contrast to the previous considerations, in which we were able to allow for changes in current arising from changes in sample temperature, for an arbitrary temperature distribution we have to limit consideration to the case of constant current. The results of the analysis may also be applied to a fairly long conductor possessing a considerable inductance, i.e., one composing, for example, part of a coil. It is also clear that when the resistance increases, the current through the sample may be reduced (for example, by using a low-resistance shunt); this can only increase the stability of the state of the sample.

The stability of any steady temperature distribution along the conductor $\tau_c(X)$ in relation to infinitely slight perturbations for any unvarying current may easily be analyzed in the following way [69]. Let us write the one-dimensional heat-conduction equation for the conductor

$$\frac{\partial \tau}{\partial t} = \frac{\partial^2 \tau}{\partial X^2} - Q(\tau) + W'(\tau), \tag{5-69}$$

where Q and W' are the dimensionless quantities of outflowing and evolving heat, which depend on the temperature.

Let us assume that as a result of fluctuations or some external action the initial steady distribution $\tau_c(X)$ is changed by a small amount $\beta(X)$. Let us substitute the new temperature distribution $\tau_c(X) + \beta(X)$ into (5-69):

$$\frac{\partial \tau_c}{\partial t} + \frac{\partial \beta}{\partial t} = \frac{\partial^2 \tau_c}{\partial X^2} + \frac{\partial^2 \beta}{\partial X^2} + W'(\tau_c + \beta) - Q(\tau_c + \beta). \tag{5-70}$$

Since β is small, we may expand the functions $W'(\tau)$ and $Q(\tau)$ in series, confining attention to the first two terms of the expansion:

$$\frac{\partial \tau_c}{\partial t} + \frac{\partial \beta}{\partial t} = \frac{\partial^2 \tau_c}{\partial X^2} + \frac{\partial^2 \beta}{\partial X^2} + W'(\tau_c) + \frac{\partial W'}{\partial \tau}\beta - \frac{\partial Q}{\partial \tau}\beta - Q(\tau_c). \tag{5-71}$$

Allowing for the fact that τ_c is a steady-state temperature distribution, we find that in the linear approximation the slight perturbation β satisfies the equation

$$\frac{\partial \beta}{\partial t} = \frac{\partial^2 \beta}{\partial X^2} - \left(\frac{dQ}{d\tau} - \frac{dW'}{d\tau}\right)\beta. \tag{5-72}$$

Let $\beta_\lambda(X)$ be an eigenfunction of the operator on the right-hand side of this equation with eigenvalues λ, i.e.,

$$\frac{d^2\beta_\lambda}{dX^2} - \left(\frac{dQ}{d\tau} - \frac{dW'}{d\tau} \right) \beta_\lambda (X) = - \lambda\beta_\lambda (X). \qquad (5\text{-}73)$$

It is well known that the functions $\beta_\lambda(X)$ form a complete orthogonal system of functions, and any arbitrary function $\beta(X)$ may be expressed in the form of a series in the functions β_λ

$$\beta(X) = \sum_\lambda a_\lambda \, \beta_\lambda (X). \qquad (5\text{-}74)$$

Let us substitute this series in into Eq. (5-72)

$$\sum_\lambda \frac{\partial a_\lambda}{\partial t} \beta_\lambda (X) = - \sum_\lambda \lambda\beta_\lambda (X) \, a_\lambda, \qquad (5\text{-}75)$$

whence

$$a_\lambda (t) = a_\lambda (0) \, e^{-\lambda t} . \qquad (5\text{-}76)$$

When eigenfunctions exist for which $\lambda < 0$, the corresponding components of the initial perturbation in the linear approximation increase without any limit, which indicates the instability of the distribution. If eigenfunctions for which $\lambda < 0$ do not exist, then all small perturbations will die out with time. Thus by studying the spectrum of the eigenvalues in Eqs. (5-73) we may in principle solve the equations as to the stability of any initial distribution. We note that this problem is completely analogous in form to the quantum problem of finding the energy spectrum of a particle in a one-dimensional potential "well." Equation (5-73) is the Schrödinger equation in which the role of potential is played by the function $[(dQ/d\tau) - (dW'/d\tau)]$, which takes both positive and negative values.

By way of a simple example, let us consider the case in which the resistance r of the conductor changes abruptly from zero to r = 1 at a critical temperature $\tau = 1 - \iota$ (see Fig. 5-2a). From the original heat-conduction equation we may derive an equation for the perturbation $\beta(X, t)$ in the form

$$\frac{\partial\beta}{\partial t} = \frac{\partial^2\beta}{\partial X^2} - \beta + \alpha\iota^2 \frac{\partial r}{\partial t} \beta. \qquad (5\text{-}77)$$

For the discontinuous function $r(\tau)$ taken, the derivative $\partial r/\partial t$ is always equal to zero except at the transition point at which it forms a Dirac δ function. In order to determine the coefficient of the δ function expressed as a function of the coordinate, we calculate the following integral in a small region close to the transition point:

$$\int \frac{\partial r}{\partial t} \, dX = \int \frac{dX}{dt} \frac{\partial r}{\partial t} \, dt = - \frac{dX}{d\tau} \bigg|_{\tau=1-\iota} . \qquad (5\text{-}78)$$

In the superconducting region r = 0 and

$$\tau(X) = (1-\iota)e^{-X+L}, \tag{5-79}$$

where L is the coordinate of the transition point. Hence

$$\tau(L) = 1-\iota. \tag{5-80}$$

Thus the potential u(X) in this problem is derived by the equation

$$u(X) = 1 - \frac{\alpha\iota^2}{1-\iota}\,\delta\,(X-L). \tag{5-81}$$

In order to find the limits of stability it is sufficient to derive the condition under which the energy of the ground state in a "well" of specified potential becomes equal to zero [we remember that u(∞) = 1]. The function β_0 of the ground state is symmetrical with respect to the origin of coordinates and (as may easily be verified) assumes the following form:

$$\beta_0(X) = \begin{cases} \cosh X & 0 < X < L \\ (\cosh L)\,e^{L-X} & X > L. \end{cases} \tag{5-82}$$

At the point X = L the derivative of the function β_0 has a break

$$\frac{d\beta_0}{dX}\bigg|_{L+} - \frac{d\beta_0}{dX}\bigg|_{L-} = \sinh L - \cosh L. \tag{5-83}$$

It is clear from Eq. (5-77) that this break or jump should equal the coefficient of the δ function in the potential u(X), since all the remaining terms of the equation remain finite, i.e.,

$$\cosh L - \sinh L = \cosh L\,\frac{\alpha\iota^2}{1-\iota}, \tag{5-84}$$

or

$$\tanh L = \frac{1-\iota-\alpha\iota^2}{1-\iota}. \tag{5-85}$$

This equation determines the position of the broken line L(ι) in Fig. 5-3b separating the regions of stable and unstable states. Since for the r(τ) relationship assumed the voltage on half the sample equals Lι, the equation of the analogous curve on the volt−ampere characteristic (Fig. 5-5a, broken line) takes the following form:

$$u(\iota) = \iota \tanh^{-1}\frac{1-\iota-\alpha\iota^2}{1-\iota} = \frac{\iota}{2}\ln\frac{2-2\iota-\alpha\iota^2}{\alpha\iota^2}. \tag{5-86}$$

In an analogous manner we may find the limits of the stability regions for more complicated cases as well (Fig. 5-2b). The corresponding derivations will not be presented here as they are very cumbersome.

On allowing for all possible additional factors (contact electrical or thermal resistance, finite nature of the resistance of the superconductor, and so on), the difficulties of calculation increase very rapidly, and an exact result can only be achieved by numerical computation. It is accordingly of special interest to present the following important result, which is of extremely general application.

In the experimental study of thermal equilibrium states in combined conductors it is frequently found that, when once a resistive section has been formed in the sample, it continues in stable existence during subsequent changes of current, even though the causes underlying its development (initial heating, flux jump, etc.) have long since vanished. This resistive or completely normal state only extends over a certain finite length, outside of which the sample may remain completely superconducting. Sometimes there are several such regions, which merge when the current changes, move spasmodically along the conductor, and so forth.

Clearly from the point of view of the normal operation of the superconducting magnetic system the possible appearance of these resistive regions is inconvenient, since even if they fail to evoke the propagation of normal regions they themselves introduce undesirable losses. It is therefore important to discover what factors might lead to the stability of such limited regions. We note that Fig. 5-3a contains a line describing such states, which corresponds to zero power of the heater ($f = 0$). However, this line lies entirely in the shaded region of unstable states. This result appears to be valid for a homogeneous conductor in the case of any unique temperature dependences of the heat evolution and heat outflow.

In order to prove this, let us consider a certain steady-state solution of Eq. (5-11) corresponding to a cupola-shaped $\tau(X)$ curve decreasing to zero at infinity (Fig. 5-1) [73]. In view of the uniformity of the conductor, any solution $\tau_1(X + \Delta X)$ also satisfies the original equation. Let us consider the function

$$\beta_1 = \frac{\partial}{\partial X} \tau_1(X) \tag{5-87}$$

and the difference between the right-hand sides of the original heat-conduction equation (5-69) for two corresponding solutions, which is equal to zero by virtue of the stationary condition

$$\frac{\partial^2 \tau}{\partial X^2}(X + \Delta X) - \frac{\partial \tau^2}{\partial X^2}(X) - Q(X + \Delta X) + Q(X) +$$

$$+ W(X + \Delta X) - W(X) = \left\{ \frac{\partial^2 \beta_1}{\partial X^2} - \left[\frac{dQ}{d\tau} - \frac{dW}{d\tau} \right] \beta_1 \right\} \Delta X = 0. \tag{5-88}$$

Thus β_1 is an eigenfunction of the operator under consideration; it corresponds to a zero value. This is very easy to interpret: the addition of perturbations of type β_1 to the original function $\tau(X)$ leads to a slight shift in the temperature distribution, which would also have remained stationary if there had been no other perturbations. However, the function β_1 has one zero. It is well known that an eigenfunction having no zeros is characterized by a smaller (i.e., negative) eigenvalue (the wave function of the ground state in quantum mechanics does not have any zeros). The assertion which we desired to prove follows from this.

According to Eq. (5-76) such a nonvanishing eigenfunction increases unlimitedly with time. This means that the initial distribution necessarily starts either "spreading" (the temperature everywhere increasing) or falling away, depending on the "content" (i.e., proportion) of the β_1 component in the original perturbation.

The theorem so proved is easily generalized to cases in which Q and W depend on the derivatives $\partial \tau / \partial X$, $\partial^2 \tau / \partial X^2$, etc., when the thermal conductivity of the conductor varies with temperature, and also to the case of moving quasi-stationary distributions. On taking account of a variety of additional effects, the basic equation (5-72) may change appreciably but it appears likely that the result will remain applicable for a very wide class of systems.

We see from the proof here presented that there is no difficulty in extending it to an arbitrary steady-state distribution in a conductor of the longitudinally homogeneous kind. If the original distribution contains several maxima, the zero eigenvalue will correspond to an eigenfunction with the same number of zeros. Correspondingly all the components of a perturbation with a smaller number of zeros are unstable. If, however, a steady-state solution exists such that the function $\tau(X)$ increases monotonically from zero to a certain finite value, then its derivative has no zeros and once again we have an eigenfunction with a zero eigenvalue. We constructed such a steady-state distribution when finding the minimum propagation current ι_p. We may accordingly conclude that such a distribution corresponds to an indifferent equilibrium, since perturbations only lead to small displacements of the original distribution, while all the more complex components of the perturbation are attenuated.

The theorem which we have thus proved clearly cannot be applied directly to the case of ambiguous $r(\tau)$ relationships. A difference between the results of a shift in the temperature distribution and differentiation is here formally possible. In this connection it was suggested in [54, 56] that such an ambiguous $r(\tau)$ relationship (of the type illustrated in Fig. 4-10) might in fact lead to the stabilization of limited normal regions. We may formally calculate the value of the current ι_p for the propagation of the normal zone by considering that the transition into the superconducting state takes place from the point on the extreme left, at which it is still possible for a resistive state to exist, as indicated by the arrow in Fig. 4-10. On the other hand, in calculating the current for the propagation of the superconducting zone we may assume that the transition into the resistive state proceeds from the point corresponding to the critical temperature (at the current specified). The values of the propagation currents thus obtained for the normal and superconducting phases should differ, i.e., there should in general be a range of currents for which neither one nor the other zone propagates.

By making assumptions of a similar kind, we may also obtain the corresponding stability criteria, which in any event give a certain lower limit for the propagation current ι_p. It is also quite possible that, for a sufficiently rapid motion of the normal or superconducting zone, the transitions between these states will take place almost exactly as we have just described. Thus, for example, we may calculate the velocity of each of the phases. However, how justified we are in introducing such a double significance of the $r(\tau)$ function when plotting the steady-state distribution (as, for example, in determining ι_p) is as yet not entirely clear.

Actually the position of the transition point in Fig. 4-10 should be determined by a more general system of equations, allowing for the tangential current components as well. Close to this point the assumption made in order to plot the $r(\tau)$ relationship (to the effect that these components are small) is no longer valid, and the effective resistance should vary continuously, although quite rapidly, from zero to the values corresponding to the upper branch of the $r(\tau)$ curve. Thus the multivalued nature of the $r(\tau)$ function may be illusory in the present situation, and then all the considerations as to the possible stable existence of limited normal regions lose their validity.

On the whole this question has at present been very little developed either theoretically or experimentally, since the separation of the individual mechanisms which might lead to the stabilization of limited resistive regions is experimentally quite difficult. It should clearly nevertheless be remembered that the detection of such regions indicates considerable nonuniformity in the properties of a particular conductor in the longitudinal direction [74].

The foregoing analysis, of course, only relates to the case of infinitely small perturbations. However, in practice the most important sources of fluctuations create finite (but fairly small) perturbations. These include jumps in magnetic flux and the losses due to the mechanical displacement of weakly pinned vortices, and variations in heat elimination associated with the formation of individual vapor bubbles. Although, in connection with the latter mechanism, cases are known in which, owing to the action of flucutations, even a very considerable supercooling of the sample is quite possible [75], on the whole the finite nature of the perturbations worsens the stability of the conductor. Quantitative estimates as to the extent of such an action are poorly defined, and for an exact revelation of the true regions of stability one must rely on experimental material.

5.6.   Experimental Results

An experimental investigation into the equilibrium states of a combined conductor for the case in which the normal and/or resistive zones exist in a steady state over a certain length of the conductor was carried out using the method of the adjustable low-resistance shunt which we employed at an earlier stage for studying the equilibrium states of a combined conductor under isothermal conditions (Section 4.7). This investigation was carried out under cooling conditions in which the combined conductor having a longitudinal temperature gradient was located in a large volume of liquid helium. As in the measurements considered earlier, the use of a shunt enabled us to realize the unstable states of the conductor experimentally.

The test sample of combined conductor was wound on a Teflon sprocket frame. In order to be able to compare experimental data with the results of calculations executed on the assumption of an infinite length of the combined conductor (Section 5.3) we used longer samples of the conductor than those used under isothermal conditions.

The temperature gradient along the sample was created by means of a permanently connected microheater placed in the center of the test sample. The rest of the experimental arrangement was the same as in the series of experiments under isothermal conditions. We recorded the following quantities: the current in the test sample, the potential difference at the ends of the sample, and the power of the microheater (from the measured values of the current in the heater circuit and the potential difference at its ends).

In the present series of experiments there was in general no fundamental necessity of measuring the voltage drop in several measuring regions along the sample as we did in the experiments under isothermal conditions. Using potential leads from the ends of the sample we measured the voltage drop over the entire sample, while the auxiliary measuring sections served to determine the disposition of the normal zone.

For auxiliary purposes we simultaneously measured the voltage drop at the ends of a short (approximately 0.5 cm) measuring section, the microheater located within this section. In addition to this, at both ends of the test sample (close to the points where it was soldered to the shunt) we monitored the voltage drop in similar short measuring sections. As soon as a potential difference appeared in these sections the measurements stopped, since its appearance indicated that the entire sample had passed into the normal state.

As in the experiments carried out under isothermal conditions, the volt−ampere characteristic of the sample was recorded with a two-coordinate automatic recorder. The total current in the circuit was gradually increased (with the power of the microheater kept constant) for a specified value of the external magnetic field. The measurements were carried out for various values of the induction between 2 and 5 T.

Our investigation into the equilibrium states of the conductor were aimed firstly at evaluating the degree of agreement between the experimental results and the calculated equilibrium states in the presence of a temperature gradient along the conductor, and secondly at determining the values of I_m and I_p so as to establish how much these differed from the corresponding values obtained earlier (Section 4.8). As the test sample for study we chose an eight-strand combined conductor without any insulation, its basic parameters being those given in Section 4.8.

A comparison between the critical currents of eight-strand samples studied under isothermal conditions and in the presence of a longitudinal temperature gradient shows that the corresponding values differ very little (usually by no more than 2-3%).

We may therefore consider that the samples to be compared have the same stability criteria α and α' as those studied under isothermal conditions.

According to the data presented in Chapter 4, for an eight-strand conductor without insulation subject to a longitudinal temperature gradient, $\alpha' > 1$ for an induction of 2, 3, and 4 T, and $\alpha' < 1$ for B = 5 T.

The results of our experiments with the uninsulated eight-strand conductor in the presence of a longitudinal temperature gradient appear in Fig. 5-12. This figure illustrates the current dependence of the voltage drop in a section of combined conductor existing in the normal (or resistive) state, and also the resistance of the normal zone for various powers of the microheater and various values of B. The broken line in the graphs indicates parts of the f = const lines not realized experimentally.

It should be noted that the results of our calculations based on the concept of a linear temperature dependence of the resistance of the combined conductor in the resistive state, allowing for the boiling crisis (Fig. 5-8), agree fairly closely with the experimental data. In actual fact the theoretical and experimental diagrams are characterized by exactly the same form of the f = const lines, the same curvature, and the existence of characteristic current values (I_m and I_p for the case in which $\alpha > 1$. For B = 5 T (for which $\alpha' < 1$) unstable states characterized by negative values of the derivative $(\partial R / \partial I)_{W_h}$ are absent; there are no values of I_m and I_p differing from I_{ct}.

As already noted, during these experiments we also measured the voltage drop at the ends of a short measuring section containing a heater. These auxiliary measurements had the following purpose.

When the current increases steadily with W_h = const, until it becomes too large, the superconductor remains in the superconducting state within this measuring section. For higher currents it assumes the resistive state. Finally, at one specific value of the current the temperature at the microheater site reaches T_{c0} and a normal zone appears in the superconductor; any further increase in the current leads to the growth of this zone.

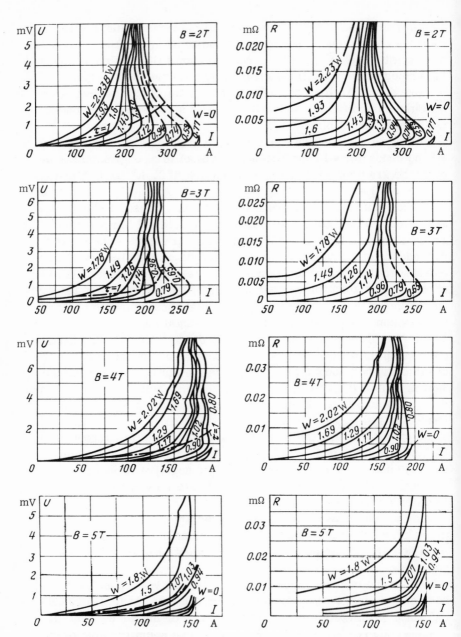

Fig. 5-12. Current dependence of the voltage drop in a section of conductor existing in the normal (resistive) state, and also of the resistance of the normal zone for various microheater powers.

Let us consider how this process is reflected in the volt — ampere characteristics recorded by reference to the auxiliary measuring section. By way of example Fig. 5-13 shows some typical relationships of this kind obtained for the eight-strand conductor. The voltage drop U in these curves is referred to 1 cm length of the conductor. While the superconductor within the measuring section is in the superconducting state, there is no voltage drop in this section. As the current increases a resistive zone is created and its temperature starts increasing; the value of U for a specified microheater power W_h = const also starts rising. After a temperature of T_{c0} has been reached (for a specific current in the measuring section) and this section has been filled with the normal zone, on further increasing the current the volt—ampere characteristic becomes linear.

Obviously the straight line corresponding to the normal state of the auxiliary measuring section will be the same for any values of W_h. The slope of this straight line (passing through the origin of coordinates) will be determined by the resistance of the substrate material. Thus the point of intersection of the line W_h = const with the straight line in question corresponds to the current

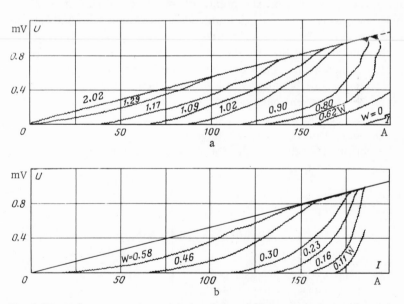

Fig. 5-13. Volt—ampere characteristics of the auxiliary measuring section. a) No insulation; b) conductor with organic insulation.

at which a normal zone appears at the site of the microheater for the specified value of W_h, i.e., the critical temperature T_c is reached. Of course the temperature T_c is reached at the site of the microheater slightly earlier, i.e., for a slightly lower current than the current at which the normal zone fills the entire auxiliary measuring section. The error arising from this effect is smaller the shorter the length of the measuring section.

Clearly by using the volt−ampere characteristic of the auxiliary section we may construct a line constituting the geometrical locus of points with $T = T_c$, i.e., $\tau = 1$ (Fig. 5-12). Comparison with Fig. 5-8 shows that the position and character of these lines obtained from the experimental data agree closely with the calculated results.

Experimental results obtained in the presence of a longitudinal temperature gradient enable us to determine the minimum current corresponding to the existence of a normal zone very simply and accurately. As indicated in Section 5.2, for $I > I_m$ the lines $f = $ const on the $L(\iota)$ diagram have a negative slope. We therefore used the following method of determining I_m for a given value of B. First by numerical methods we determined the values of the derivative $(\partial I/\partial R)_{W_h}$ at the points of inflection of the lines $W_h = $ const (the derivative has its smallest value for a specified W_h at these points). These values were plotted on the $(\partial I/\partial R)_{W_h} = f(I_{pi})$ graph, where I_{pi} is the current corresponding to the points of inflection in question. This curve was only plotted instead of $(\partial R/\partial I)_{W_h} = f(I_{pi})$ because for $I = I_m$ we have $(\partial R/\partial I)_{W_h} = \infty$ and in this case the determination of I_m is much less accurate than in the method described here.

Of course for $I_{pi} > I_m$ we have $(\partial I/\partial R)_{W_h} < 0$ while for $I_{pi} < I_m$ we obtain $(\partial I/\partial R)_{W_h} > 0$. By joining the points on the graph with

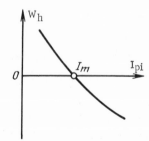

Fig. 5-14. Curve for determining the minimum current corresponding to the existence of a normal phase I_m.

TABLE 5-1

B, T	Experimental data					Calc. of ι_m by Eq. (5-89)	$\delta\iota_{m'}$ %
	I_{ct}, A	I_m, A	$\iota_m=I_m/I_{ct}$	I_p, A	$\iota_p=I_p/I_{ct}$		
2	366	200	0.545	224±10	0.61	0.54	0.9
3	259	198	0.76	210±7	0.81	0.74	2.6
4	184	170	0.92	175±5	0.95	0.92	0

a smoothing curve and finding the intersection of this curve with the horizontal axis we obtain the value of I_m for the specified value of B (Fig. 5-14). The values of I_m for the eight-strand conductor in the case $\alpha' > 1$ are presented in Table 5-1. The same table indicates the minimum propagation current I_p found by means of the experimental R(I) curves. In determining this current we used the fact that I_p formed the limit to which all the W_h = const lines tended on the R(I) diagram for a conductor with a longitudinal temperature gradient.

Of course this method of determining I_p should be regarded as fairly approximate. We see from Table 5-1 that the possible error in determining I_p may (according to our own data) rise as far as ±4.5%. However, this method of determining I_p is perfectly satisfactory for various kinds of rough estimates.

It is of great interest to compare the results of our determination of I_m from experimental data with the results of an approximate calculation based on the known value of I_p and carried out in the following way. From (4-28) and (5-6)

$$\iota_p = \frac{I_p}{I_{ct}} = \sqrt{\frac{1}{4\alpha^2} + \frac{2}{\alpha}} - \frac{1}{2\alpha}$$

we derive a simple relationship between ι_m and ι_p:

$$\iota_m = \iota_p/\sqrt{2 - \iota_p}. \tag{5-89}$$

Table 5-1 shows the values of ι_m obtained experimentally and the calculated values obtained from Eq. (5-89) for the experimental values of ι_p. The comparison reveals excellent agreement between the experimental and calculated values of ι_m; the difference, never more than 2.6%, nowhere exceeds the limits of error corresponding to the determination of I_p.

TABLE 5-2

B, T	I_p, A	$\tilde{I}_p$, A
2	224±10	275
3	210±7	240
4	175±5	220

It is also especially interesting to compare the experimental values of I_p from Table 5-1 with the values of the current $\tilde{I}_p$ at which the normal zone vanishes in a sample subjected to isothermal conditions, as given in Chapter 4 (Fig. 4-41). The values of $\tilde{I}_p$ are everywhere much higher than I_p (Table 5-2).

The difference between I_p and $\tilde{I}_p$ may be explained in the following way. Let us consider the sequence of states in a sample subjected to isothermal conditions for cases in which the sample is initially in the normal state and the current passing through it is reduced. The corresponding volt–ampere characteristic is shown in Fig. 5-15.

When the current is reduced the sample under consideration passes from the state corresponding to point a into the state corresponding to point B. On reaching point B, for which the current $I' > I_p$, the normal zone starts diminishing. From this instant the sample no longer exists under isothermal conditions but has a longitudinal temperature gradient; the experimentally determined line $B-C$ forms part of the line $W_h = 0$. The point B is the point of intersection of the isotherm $T = T_c$ with the line $W_h = 0$. Since the current I_p is the limit to which the line $W_h = 0$ tends for an unlimited rise in U, it is clear that the current I' corresponding to the transition from the isothermal situation to that of the longitudinal temperature gradient is always greater than the propagation current I_p. The difference $I' - I_p$ is smaller the steeper the $T = T_{s0}$ isotherm, i.e., the longer the test sample. It should be emphasized that the line $B-C$ corresponds to a reduction of the total current in the shunt–sample system, although the current in the sample itself rises.

The collapse of the normal zone occurs for a current $\tilde{I}_p$ at point C, at which the derivative $(\partial U / \partial I)_{W_h=0}$ becomes equal to $(\partial U / \partial I_{sh})_{I_{tot}}$ (line $m-C-n$ in Fig. 5-15).

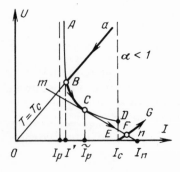

Fig. 5-15. Volt–ampere characteristics of a sample in the normal state under isothermal conditions. OBa, $T = T_c$ isotherm; EFG, line $\alpha = $ const $(\alpha < 1)$; $ABCD$, line $W = 0$ in the presence of a longitudinal temperature gradient.

Hence the difference between the values of $\tilde{I}_p$ and I', and correspondingly the difference between $\tilde{I}_p$ and I_p, for the same length of sample will be the smaller the sharper the shunt characteristic. However, as already noted in Chapter 4, in order to be able to study the unstable states completely this characteristic should be as flat as possible; under these conditions a considerable difference between the values of $\tilde{I}_p$ and I_p is inevitable.

We ought to consider one further problem — into what state does the sample pass after the collapse of the normal zone has taken place at point C? The answer to this question may be obtained from Fig. 5-15. If the shunt characteristic is so steep that the total current I_{tot} corresponding to the point C is smaller than I_{ct}, then after the collapse of the normal zone the sample will pass into the superconducting state. If, however, the shunt characteristic is not too steep and $I_{tot} > I_c$, then this characteristic will intersect the line α = const in the region of the resistive state, and thus after the collapse of the normal zone the sample will become resistive. This is the situation depicted in Fig. 5-15, where the sample passes from point C to point F.

Using the microheater we carried out a special series of experiments of interest from the point of view of analyzing the conditions associated with the development of the liquid helium boiling crisis on the surface of the conductor. These experiments amounted to the following. For a fixed value of the current in the sample I, the microheater power was increased from zero to a value W_h at which a potential difference of $\Delta U = 1\ \mu V$ appeared in the auxiliary measuring section just described. The appearance of this potential difference indicated that for this value of W_h the superconductor had passed into the resistive state. By repeating this procedure for various I values we may determine the W_h (I) relationship.

The higher the value of the current I, the lower will be the sample temperature and hence the lower the microheater power W_h at which the resistive zone appears. Clearly for $I = I_c$ the power $W_h = 0$, while W_h will reach its maximum when I = 0. The transition to the resistive state will occur when the conductor is heated to such a temperature that the specified current I is critical. Thus with each value of the current at the transition point we may correlate a temperature determined by means of Eq. (4-3):

$$T = (T_{co} - T_t)\left(1 - \frac{I}{I_{ct}}\right) + T_t, \qquad (5\text{-}90)$$

or, in dimensionless notation,

$$\tau = 1 - \iota. \qquad (5\text{-}91)$$

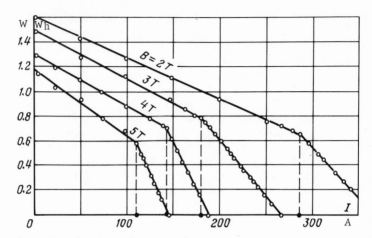

Fig. 5-16. The $W_h(I)$ relationship for an eight-strand conductor without insulation.

The current $I = I_{ct}$ corresponds to the value $T = T_t$ and the current $I = 0$ to $T = T_{c0}$.

If the heat-transfer coefficient varied monotonically with temperature over the whole temperature range of the conductor from $T = T_t$ to $T = T_{c0}$, the $W_h(I)$ relationship would have a smooth characteristic. The existence of a boiling crisis leads us to expect a break to appear on this characteristic curve at $T = T_M$, since in the range T_t to T_M the heat release from the sample surface is greater (bubble type or boiling) than in the range T_M to T_{c0}.

The results of our experiments for the eight-strand conductor are presented in Fig. 5-16. We see from this figure that the boiling crisis causes a sharp break in the $W_h(I)$ curve; to the left and right of the break the relationship is linear. For comparison, Fig. 5-17 shows the results of an analogous experiment with an eight-strand conductor having organic insulation, the parameters being given in Section 4.8. As already mentioned, as a result of the existence of thermal resistance due to the insulation no boiling crisis appears on the volt—ampere characteristics (see Fig. 4-40). This cannot fail to be reflected in the characteristic of the $W_h(I)$ relationship. We see from Fig. 5-17 that there are no breaks in the curves.

On analyzing the experimental data presented in Fig. 5-12 it is not difficult to see that for cases in which $\alpha' > 1$ the range of states close to $I = I_{ct}$ has had little study; the diagrams have no $W = \text{const}$ lines closely approaching the critical point. This is be-

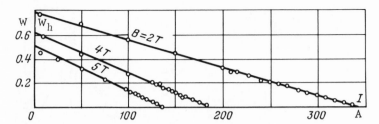

Fig. 5-17. The $W_h(I)$ relationship for an eight-strand conductor with organic insulation.

cause close to $I = I_{ct}$ and for low values of the microheater power W_h, the length of the section of combined conductor occupied by the normal zone becomes so great that it exceeds the dimensions of the sample under examination (the length of the sample in our experiments was never greater than a few tens of centimeters). This prevented us from making experiments for low W_h values close to the critical current; in order to carry out such an experiment we would have had to place a much longer wire in a zone of as uniform a field as possible.

The region close to the critical currents was studied in detail in a series of special experiments with a long sample, also made of uninsulated eight-strand conductors. These experiments also had another aim: that of studying the volt−ampere characteristics of a combined conductor in the presence of a longitudinal temperature gradient, not under the conditions of heat transfer associated with a large volume of liquid helium but under the conditions of heat transfer usually encountered for a solenoid winding.

The experiment was arranged as follows. On a Teflon coil-frame we wound nine or ten layers of a Manganin wire coil with the same diameter as the combined conductor under consideration. Then on top of this coil, in a single layer of about 100 turns, we placed a bifilar winding of the test sample (combined conductor), the length of the winding being 15-20 m. A microheater was fixed to the sample. Above this layer we again wound several layers of Manganin wire. The channels for the liquid helium, as is the case in the winding of superconducting solenoids, were created by winding a Capron filament on the conductor.

Thus, from the point of view of the mode of heat transfer, the test sample existed under conditions as closely as possible ap-

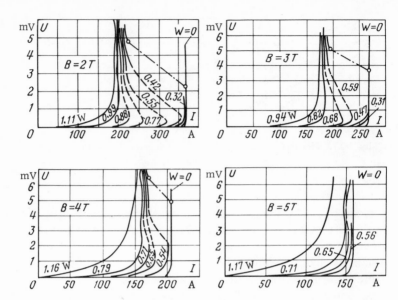

Fig. 5-18. Volt—ampere characteristics of a long sample of an eight-strand conductor.

proaching those used for cooling the inner layers of the winding of a superconducting solenoid. The great length of the sample enabled us to study the range of states close to the critical current. The sample was wound in the bifilar manner so as to reduce the inductance of this layer of the coil to zero: In this case the measured potential difference at the ends of the sample equaled the active voltage drop in the latter. As in the previous case, the ends of the sample were soldered to a low-resistance shunt. The external magnetic field was created by means of a superconducting solenoid with an internal diameter of 90 mm.

Figure 5-18 shows the $U(I)_{W_h = \text{const}}$ relationship obtained for a long sample of an eight-strand conductor with induction of 2, 3, 4, and 5 T. It should first of all be noted that no serious influence of any changes in the conditions of heat transfer from the surface of the combined conductor to the boiling liquid helium on the character of the $U(I)_{W_h = \text{const}}$ relationships was established as a result of these experiments. In the experiments we made a detailed study of the range of states close to the critical current for a combined conductor subjected to a temperature gradient.

The results agree closely with the results of a calculation carried out with due allowance for the boiling crisis (see Fig. 5-8).

The principal conclusion emerging from the resultant data is as follows: It confirmed that for a number of values $W_h \geq 0$, the voltage drop in a combined conductor with a longitudinal temperature gradient for $I = I_{ct}$ has a very substantial value. This conclusion is based on the results of experiments in which we were able to attain parts of the $W_h = 0$ lines in the region close to I_p.

The experiments in which the states $W_h = 0$ were reached were carried out in the following way. For a prespecified, fairly low microheater power we gradually increased the current in the shunt–sample system; we recorded the volt–ampere characteristic represented by the line $a-b$ in Fig. 5-19. After the current in the sample had been reduced to a value close to I_p (point b), the microheater was disconnected and the sample passed into the state corresponding to $W_h = 0$ (point c).

Then the total current in the shunt–sample system was reduced, whereupon the current in the sample increased and the volt–ampere characteristic corresponding to the line $W_h = 0$ $(c-d)$ was recorded.

On reaching point c, determined by the condition $(\partial U/\partial I)_{W_h} = (\partial U/\partial I_{sh})_{I_{tot}}$, the normal (or resistive) zone in the sample loses stability and vanishes. The next stable state into which the sample passes lies close to the intersection of the line $I_c = $ const and the shunt characteristic $(c-e)$; the state corresponding to the point e is uniquely established by experiment. It follows from Fig. 5-19 that the point corresponding to the voltage drop in the sample for $W_h = 0$ and $I = I_{ct}$ lies above the point e. Since the voltage drop in the sample in the state corresponding to point e is quite large, so is the voltage drop for $W_h = 0$ and $I = I_{ct}$.

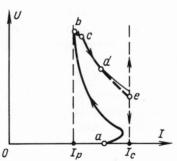

Fig. 5-19. Volt–ampere characteristic of the sample in experiments carried out to determine the W = 0 state.

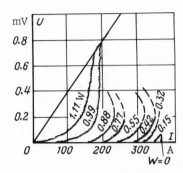

Fig. 5-20. Volt—ampere characteristics of the auxiliary measuring section for B = 2 T.

After the sample had passed into the state corresponding to
the point e, varying the total current in the shunt—sample system
hardly affected the current in the sample, with the latter remaining approximately equal to the critical current; the voltage in the
sample thereupon varies in accordance with the changes in total
current (see Fig. 5-18).

The results of this series of experiments also confirm the
second characteristic feature of the $U(I)_{W_h = const}$ diagram, and
that is the presence of a break on the W_h = const lines at points
corresponding to a temperature near the onset of the boiling crisis
$(T = T_M)$. On increasing the total current in the shunt—sample
system, the current in the sample increases until the temperature
close to the microheater reaches T_M. Shortly after this, the current in the sample changes abruptly, and the voltage drop is sharply increased. This abrupt change is due to the boiling crisis, as
confirmed by measurements of the voltage drop in the auxiliary
measuring section. The results of such measurements for the
case B = 2 T are presented in Fig. 5-20. We see from the figure
that for values of W_h smaller than a certain level there is a characteristic break (similar to that described in Chapter 4) on the
W_h = const line in the region of the resistive states; this corresponds to the transition into the normal state arising from the
boiling crisis.

These experimental results fully confirm the validity of the
model representations which we developed for combined conductors
with a longitudinal temperature gradient, allowing for the boiling
crisis in liquid helium.

Chapter 6

Propagation of the Normal Zone in a Superconducting Coil

6.1. Method of Studying the Propagation of the Normal Zone

In order to carry out a complex analysis of the laws governing the development, existence, and propagation (expansion) of the normal zone in the coil of a superconducting magnetic system, apart from studying the static conditions of equilibrium it is also essential to study the dynamic processes involved in the extension and contraction of the normal zone in a combined superconductor. Since, in the course of its extension, the normal zone embraces a much greater length of the conductor than under steady-state conditions, in conducting these experiments we chose a combined conductor 200-300 m long, forming a solenoid.

It should be noted that the transition from a sample situated in a large volume of liquid helium to a sample constituting a loose winding* naturally leads to a reduction in the heat-transfer coefficient from the surface of the combined conductor, and hence to a reduction in the degree of stabilization. Hence a quantitative comparison between the results of experiments carried out under such very different conditions of heat transfer is of special interest

---

*We label a winding which contains channels for the passage of the liquid helium into the inner layers "loose" to distinguish it from a compact or dense winding in which the turns lie close to one another and the penetration of liquid helium to the inner layers of the winding is almost entirely prevented.

from the point of view of determining the relationship between the heat-transfer coefficients under these conditions.

The results of our investigations into the dynamic processes involved (which will be presented in this chapter) were obtained under extremely "pure," idealized conditions, in which the sample under consideration was made in the form of a short-circuited coil. In this case the solenoid was not influenced by any perturbations in the external section of the electrical circuit; perturbation due to fluctuations of the supply voltage and interference in the regulating system were entirely excluded.

The normal zone in the test coil was created by means of a microheater placed on one of the turns inside the solenoid winding. The microheater (its initial power being selected empirically) was switched on for a short time, just as long as was necessary to create a normal zone in a particular part of the winding (about 0.1 sec). Thus the original cause of the normal zone's appearance was eliminated and the zone was left "on its own." This mode of operating the microheater simulates the formation of a normal zone as the result of a "catastrophic" flux jump or as a result of the displacement of the coil windings. After the disconnection of the microheater the subsequent behavior of the normal zone was duly recorded.

When studying the dynamic processes with a short-circuited solenoid we employed two distinct methods: "single-solenoid" and "double-solenoid." In using the first method the sample was wound in the form of an ordinary solenoid, which was then used for those experiments performed in the state of "frozen" magnetic flux; i.e., the solenoid was short-circuited with a superconducting shunt and disconnected from the supply source.

During the experiments quantities proportional to the current I in the solenoid winding and the voltage drop U_{nz} in the normal zone were recorded at each instant of time. The current in the winding of the short-circuited solenoid was determined by reference to the readings of a bismuth magnetoresistive magnetic-field sensor, previously graduated by reference to the coil current. The voltage drop in the normal zone was determined by means of a measuring coil placed inside the solenoid and inductively coupled to the latter. The following equation governs the voltage drop in question:

$$U_{nz} = -(L - L_{nz})\frac{di}{dt},\qquad (6\text{-}1)$$

$$\frac{di}{dt} = -\frac{e}{M}, \tag{6-2}$$

where e is the mutual-induction emf in the measuring coil, M is the mutual inductance of the measuring coil and solenoid, and L and L_{nz} are the inductances of the solenoid and that part of the winding which has passed into the normal state.

Careful estimates showed that, under the conditions of the experiments described here, in the majority of cases $L_{nz} \ll L$ and so

$$U_{nz} \approx iR_{nz} = -L\frac{di}{dt}. \tag{6-3}$$

On the basis of these experimental data we calculated the values of i and u_{nz} by reference to the previously determined calibration curve of the bismuth sensor and the values of L and M. Using these results, it was easy to determine the resistances of the normal zone for every value of i

$$R_{nz} = U_{nz}/i. \tag{6-4}$$

Apart from the foregoing parameters, in the course of the experiments we measured the voltage drop at the ends of a small auxiliary measuring section (roughly 1 cm long) having a microheater inducing a normal zone placed in its center.

In the experiments with the single solenoid the magnetic field was created by the solenoid itself. Thus, in the course of the experiments, when the coil current decreased as a result of the creation of the normal zone, the magnetic field surrounding the test solenoid also diminished. This, of course, made it difficult to relate the results of the experiment to the same fixed value of the field induction. The situation was further complicated by the fact that, in this form of experiment, different parts of the solenoid were subject to different magnetic fields at every specific instant (since in any solenoid the induction of its own field has a particularly sharp gradient inside the actual coil). In recounting the shortcomings of the single-solenoid method we should also mention that a strict determination of U_{nz} from Eq. (6-1) demands a knowledge of the inductance L_{nz}, a quantity which is difficult to determine accurately.

All these deficiencies seriously undermine the usefulness of the experimental results obtained with a single solenoid from the

point of view of analyzing the dynamic processes for B = const
and generalizing the resultant laws. This is why we decided on the
other experimental method which is about to be described, that of
the double solenoid, a method entirely free from the disadvantages
listed. It should be emphasized that the use of the double-solenoid
method does not avoid the necessity of using the single-solenoid
method as well. The point is that the winding of real superconducting magnetic systems operate under the same conditions as a
single solenoid, i.e., in their own magnetic field, subject to its
very considerable gradient. From this point of view the experiments carried out with the double solenoid cannot act as a substitute for experiments with a single solenoid, in which the specific
characteristics of the dynamic processes associated with practical
superconducting coils are naturally realized. Clearly, an analysis
based on the results of a "pure" experiment carried out by the
double-solenoid method should be supplemented by a study of the
dynamic processes taking place in the single-solenoid system
under conditions similar to those encountered in practical service.

The essence of the double-solenoid method is as follows [66].
The winding of solenoid 1, constituting the test sample, is executed
in the bifilar manner (Fig. 6-1). Thus the winding has no inductance
and does not create any intrinsic magnetic field. By means of a
Capron filament, liquid-helium channels are created in the winding.
At one of the internal points a microheater is attached to the coil.
Solenoid 1 is placed inside another superconducting solenoid 2,
which is used to create a fairly uniform magnetic field.

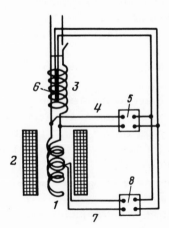

Fig. 6-1. Arrangement for measurements based on the double-
solenoid method.

The test solenoid 1 is connected in series with an ordinary superconducting solenoid 3 placed outside solenoid 2. Solenoid 3, storing a certain amount of energy, acts as a special kind of supply source for solenoid 1 when the "solenoid 1 + solenoid 3" system is in the "frozen" magnetic-flux mode. For measuring the potential difference in the test solenoid the conductors 4 are connected to the two-coordinate automatic recorder 5. Since the winding of the solenoid 1 is made in the bifilar manner, the measured potential difference is equal to the voltage drop in the active impedance of the normal zone. The second input of the two-coordinate automatic recorder receives a signal from the bismuth magnetoresistive sensor 6 placed inside solenoid 3. This signal is proportional to the current in the "solenoid 1 + solenoid 3" system. Thus the automatic recorder plots the $U_{nz}(I)$ relationship.

As in the single solenoid, potential leads 7 are soldered to the coil on both sides of the microheater at distances approximately 1 cm from one another. The potential difference from this auxiliary measuring section U_{meas} is applied to the input of another two-coordinate automatic recorder 8. The signal from the bismuth sensor 6 is also applied to the second input of this recorder, which accordingly traces the $U_{meas}(I)$ relationship.

Thus, the essence of our experimental method amounts to a measurement of the $U_{nz}(I)$ and $U_{meas}(I)$ relationships during the propagation of a normal zone (introduced into one of the points of the coil by the brief application of the diagnostic microheater) through the winding. Analysis of the experimental data provides extensive information regarding the basic laws of propagation of the normal zone through the coil of a superconducting magnetic system.

6.2. Propagation of the Normal Zone in a Thinly Packed Coil

The propagation of the normal zone in a thinly packed ("loose") coil of superconducting material was studied by the double-solenoid method in magnetic inductions of 2, 3, 4, and 5 T.

The measurements were carried out for various values of the "frozen" current larger than the propagation current I_p. Clearly for $I < I_p$ the normal zone introduced into the winding starts contracting immediately after the microheater has been disconnected.

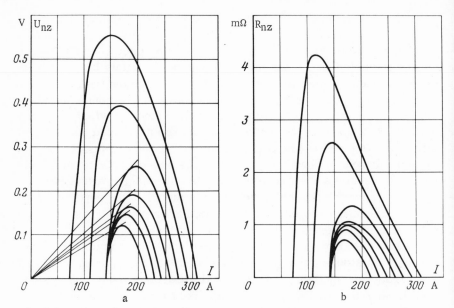

Fig. 6-2. Voltage drop in the normal zone (a) and normal-zone resistance (b) as functions of current for B = 2 T.

For $I > I_p$ the normal zone, once created, starts expanding and the resistance of the normal zone naturally increases. However, as the normal zone expands it dissipates energy stored in the "solenoid 1 + solenoid 3" system. As a result of this dissipation the current in the system falls below $I = I_p$, the zone starts contracting, and at a certain current it vanishes altogether. After this the current in the "solenoid 1 + solenoid 3" system (we may call this the residual current) remains constant.

Figure 6-2a shows the U_{nz} (I) relationship for the case in which the solenoids 1 and 3 (Fig. 6-1) were made from an eight-strand conductor without any insulation. The solenoid with the bifilar winding had an internal diameter of 30 mm, an external diameter of 80 mm, a height of 40 mm, and 346 turns in the coil; the solenoid forming the supply source had corresponding dimensions 30, 60, and 70 mm and 363 turns.

Figure 6-2b shows the R_{nz} (I) relationship; the value of R_{nz} was determined from Eq. (6-4).

The curves presented in Fig. 6-2 correspond to the development of the dynamic process after the appearance of a normal zone

in the combined conductor for various values of the initial current
in the solenoid I_0. From the general discussion just presented we
may draw the following conclusion: The vanishing ("collapse") of
the normal zone in the dynamic process here described should occur at a current equal to the minimum current for the existence of
the normal zone I_m, regardless of any dependence on the initial
current in the solenoid. In principle, for an initial current I_0 slightly exceeding the minimum normal-zone propagation current I_p,
this zone may vanish for current values lying in the range I_m to I_p.
With increasing I_0 the residual current rapidly approaches the
minimum current required for the existence of a normal zone (I_m).
The foregoing a priori conclusion is excellently confirmed by experimental data — within a specific range of initial currents (from
I_m to a certain higher value, the physical meaning of which will be
considered a little later) all the lines of the dynamic processes end
in exactly the same point on the horizontal axis. This point should
be interpreted as the minimum current for the existence of the
normal zone I_m.

It is obvious that the resistance R_{nz} on a particular curve of
the dynamic process has a maximum value at a current equal to
the normal-zone propagation current I_p. Thus, in the $R_{nz}(I)$ diagram the maxima of the various dynamic-process curves should
lie on a single vertical $I = I_p$. Correspondingly on the $U_{nz}(I)$ diagram the points at which the curves of the dynamic processes
touch the straight lines drawn from the origin should also lie on
the verticals $I = I_p$, since

$$R_{nz}^{max} = \left(\frac{U_{nz}}{I} \right)_{I_0}^{max} . \tag{6-5}$$

We see from Fig. 6-2 that over a certain range of initial currents these laws are quite clearly evident, and the I_p value may be
established without any difficulty from the diagram. However, on
further increasing the initial current the abscissas of the maxima
on the $R_{nz}(I)$ curves start deviating from the $I = I_p$ value.

This effect may be explained as follows. Let us consider Fig.
6-3, which gives the experimental $U_{meas}(I)$ curves for various
values of I_0. Here U_{meas} is the potential difference at the ends of
the short auxiliary measuring section, the ends of which lie on
both sides of the site of the diagnostic microheater; using this

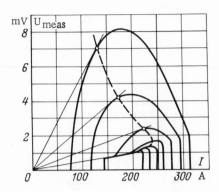

Fig. 6-3. Volt—ampere characteristics of the auxiliary measuring section for B = 2 T.

heater a normal-zone nucleus is introduced into the winding. Clearly, after the normal zone has extended a long way outside the limits
of the measuring section, the U_{meas} (I) relationship may be regarded as the volt—ampere characteristic for a section of the combined conductor which exists under practically isothermal conditions at each particular instant. The ratio U/I is the resistance
of unit length of the combined conductor at each specific moment
of time.

Since the length of this auxiliary measuring section is extremely short (about 1 cm), when the microheater is connected, the filling of the measuring section with normal-zone material is hardly
noted in the solenoid current — that is, during this process there is
hardly any diminishment in the current (see schematic diagram
in Fig. 6-4a, section $a-b$).

The greater the initial current at which the normal zone is
introduced into the coil, the greater is the intensity of the Joule
heating in this zone (for the same dimensions of the normal zone
at a specified instant of time), and hence the greater is the temperature of the combined conductor in the normal-zone section.
If the resistivity of the material composing the substrate did not
alter with temperature, all the lines of the dynamic processes
corresponding to different values of the initial current would merge
with each other on the U_{meas} (I) diagram. This would happen because in the normal state (section $b-c$ in Fig. 6-4a) the resistance
of the substrate material and hence the ratio $(U_{meas}/I)_{c0}$ would be
the same for any initial currents (i.e., for any temperatures of the
combined conductor $T > T_c$). The "collapse" of the normal zone (section $c-d$) would then occur, as already mentioned, for the same value

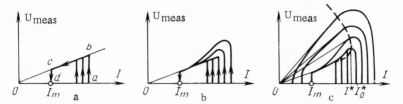

Fig. 6-4. Volt−ampere characteristics of the dynamic processes for different values of the initial current.

of the current $I = I_m$. Since the substrate material actually has a resistance which varies considerably with temperature, current effects similar to those considered in Section 4.5 start appearing in the coil under consideration. The section corresponding to the decrease in current on the curve of the dynamic process is then not rectilinear (as in the case just considered, in which the temperature range is so narrow that the temperature dependence of the substrate resistivity hardly appears at all) but has a variable slope. As I diminishes the ratio $(U_{meas}/I)I_0$ does likewise, approaching $(U/I)I_0$, $T = T_{c0}$ (see Figs. 6-3 and 6-4b).

It is very clear that the point on the line of the dynamic process (corresponding to a specified value of the initial current I_0) at which the resistance of the section of combined conductor $(U/I)I_0$ has a maximum value is the point corresponding to the equilibrium state. In actual fact the maximum resistance per unit length of the combined material capable of being reached in the dynamic process corresponds to the maximum temperature of the conductor in this process. At the point in question the sign of the derivative dT/dt changes, the temperature of the conductor in the auxiliary measuring section first rises, passes through a maximum, and then decreases.

Thus if in the U_{meas} (I) diagram we draw tangents from the origin to each of the lines representing the dynamic process we may find the points belonging to the line of equilibrium states for the particular combined conductor. In this way we may plot the whole line of equilibrium states (the line α = const) in the temperature range $T > T_c$, including the unstable branch of this line corresponding to $T > T^*$. We see from Fig. 6-3 that the character of this line agrees entirely with the results obtained in Chapter 4 by both calculation and direct experiment. In particular the maximum equilibrium current I^* is clearly established.

The fact that the experimental data regarding the dynamic processes in the short-circuited winding agree closely with our conclusions as to the character of the line of equilibrium states of the combined conductor for $T \geq T_c$ (obtained both theoretically and experimentally under isothermal conditions) is an important proof of the validity of the analysis conducted in Section 4.5.

Let us use I_0^* to denote the initial current for which the equilibrium state in the dynamic process corresponds to the maximum equilibrium current I^*. We see from Fig. 6-4 that the line corresponding to the current I_0^* forms a special kind of boundary curve; for initial currents $I_0 > I_0^*$ the dynamic process terminates, i.e., the normal zone vanishes for currents $I < I_m$. The greater the current I_0, the smaller is the residual current in the solenoid. For a particular value of I_0 the process develops in such a way that the current in the solenoid attenuates completely (Fig. 6-4b).

It is very plain from the foregoing analysis that for dynamic processes corresponding to $I_0 > I_0^*$ the concepts of the minimum current for the existence of a normal zone I_m and the minimum normal-zone propagation current I_p lose their meaning. This is

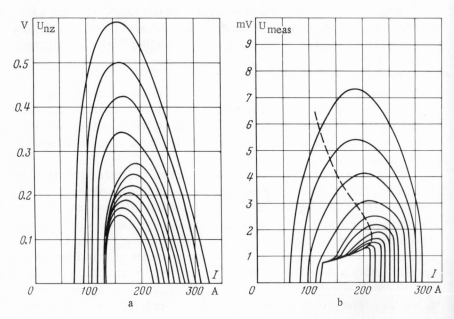

Fig. 6-5. Volt−ampere characteristics of dynamic processes for B = 3 T.

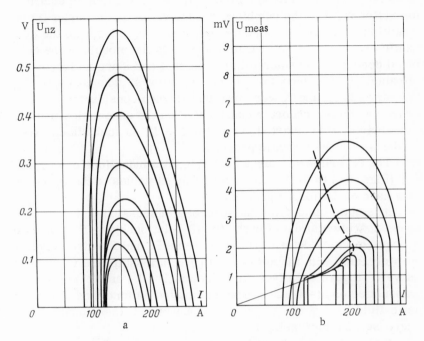

Fig. 6-6. Volt—ampere characteristics of dynamic processes for B = 4 T.

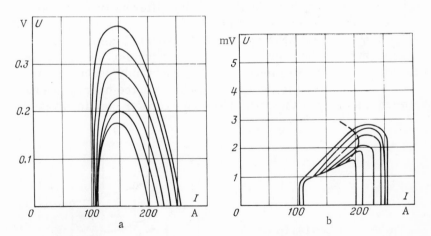

Fig. 6-7. Volt—ampere characteristics of dynamic processes for B = 5 T.

associated with the fact established in Chapter 4 that no equilibrium states exist in the combined conductor for I > I* under the
condition I = const. In other words, in order to describe the dynamic processes in this range of currents the models of the combined conductor considered in Chapters 4 and 5 are no longer applicable. The reasons for this are as follows. The concepts of
the currents I_m and I_p were introduced when considering models
of the combined conductor constructed on the assumption that stable
equilibrium states existed, while for $I_0 > I_0^*$ no such states exist
on the line of the dynamic process.

The experimental curves for magnetic inductions of 3, 4, and
5 T (Figs. 6-5 to 6-7) plotted for different values of the initial current have an analogous character.

It is quite clear from the foregoing discussion that, if we possess experimental data for the U_{nz} (I) relationship in the dynamic
process, we may determine the currents I_m and I_p extremely accurately, and, with the aid of the U(I) relationship, the value of I*
as well. Figure 6-8 gives the values of I_m, I_p, and I* for the
uninsulated eight-strand conductor studied as functions of the magnetic-field induction. As usual (see Chapter 4) these quantities
decrease with increasing B.

It should be noted that in the present series of experiments
the conditions of heat transfer from the surface of the combined
conductor to the liquid helium were very different from the heat-
transfer conditions in the earlier experiments, in which the equilibrium of the normal zone was studied. Whereas before heat transfer took place in the presence of a large volume of boiling liquid
helium, under the conditions of the loose or thinly packed coils
heat was transferred to the helium situated in the narrow channels
between the turns of the coil. The heat-transfer coefficient is of

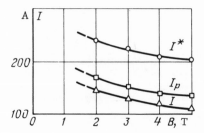

Fig. 6-8. Currents I_m, I_p, and I*
as functions of the induction of the
external magnetic field.

TABLE 6-1

B, T	I_m^{lv}, A	I_m^{coil}, A	I_p^{lv}, A	I_p^{coil}, A	I^{*lv}, A	I^{*coil}, A
2	200	144	224 ± 10	165	318	240
3	198	131	210 ± 7	151	288	226
4	170	120	175 ± 5	139	268	211

course much smaller under these conditions. The characteristic currents I_m, I_p, and I^* should decrease accordingly.

For comparison, Table 6-1 gives the values of these currents obtained in experiments involving heat transfer to a large volume of liquid helium (lv) and in the experiments with the loose coil (coil). The data presented in this table confirm that the transition to the conditions of heat transfer characteristic of the loose coil leads to a considerable reduction in the values of I_m, I_p, and I^*.

These data also enable us to estimate the relationship between the heat-transfer coefficients representing the passage of heat from the surface of the combined conductor to liquid helium occupying a large volume (h_{lv}) and to helium in the narrow channels of a loose coil (h_{coil}).

It is clear from Eq. (4-108), i.e., the equation defining the maximum equilibrium current, which on allowing for the equation $q = h(T - T_0)$ may be written

$$I^* = \sqrt{\frac{APh}{d\rho/dT}},\tag{6-6}$$

that, other conditions being equal,

$$\frac{h_{lv}}{h_{coil}} = \left(\frac{I^{*lv}}{I^{*coil}}\right)^2.\tag{6-7}$$

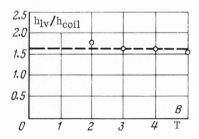

Fig. 6-9. Ratio of the heat-transfer coefficients h_{lv}/h_{coil} as a function of the induction of the external magnetic field.

Of course the ratio h_{lv}/h_{coil} should not depend on the induction B. The values of this ratio calculated by means of Eq. (6-7) from experimental data relating to I* for different values of B are presented in Fig. 6-9. We see from the diagram that the values of h_{lv}/h_{coil} are closely grouped around a mean of approximately 1.62.

Thus, in the particular case under consideration, on passing from a sample situated in a large volume of boiling liquid to a loose coil the heat-transfer coefficient from the surface of the combined conductor decreases by about 40%.

For estimating h_{lv}/h_{coil} there is also another possible method. It follows from Eq. (4-28), which gives the minimum current corresponding to the existence of the normal zone in the Stekly model, that on allowing for the stability criterion expressed in the form $\alpha = \rho I_c^2 / AhP(T_{c0} - T_t)$,

$$\frac{h_{lv}}{h_{coil}} = \left(\frac{I_m^{lv}}{I_m^{coil}} \right)^2 . \tag{6-8}$$

Structurally this equation is similar to Eq. (6-7); however, since Eq. (4-28) is only valid for combined conductors corresponding to the Stekly model, under the conditions corresponding to the bubble-type boiling of liquid helium this equation is not of a general character, in contrast to Eq. (6-7), which is not based on any particular model.

As already noted, the experimental U(I) relationships for the dynamic processes associated with various values of the initial current I_0 enable us to determine the position of the line of equilibrium states of the combined conductor. Earlier the position of this line was determined by means of the volt−ampere characteristics of the conductor obtained for the equilibrium states under isothermal conditions. It is important to realize that in the experiments carried out under equilibrium isothermal conditions (in which the test sample is situated in a large volume of liquid helium) the line of equilibrium states obtained refers to the case in which the heat passes directly from the sample surface into a large volume of boiling helium. On using the method of determining the line of equilibrium states set out in this section, the position of the line so determined refers to the conditions of heat transfer characteristic of a loose coil.

Our study of the dynamic process developing in the coil after introducing a nucleus of the normal zone, carried out by the single-solenoid method, reveals certain specific characteristics typical of the coils in real superconducting magnetic systems.

These characteristics are determined by the fact that in a single solenoid the magnetic induction is uniquely related to the current "frozen" in the solenoid. Thus for each value of the initial current I_0 we have a specific value of the induction in the solenoid at the instant of introduction of the normal zone nucleus. As the current varies in the solenoid during the dynamic process under consideration the magnetic field also varies. Thus different lines of the dynamic process in this case refer to different field inductions. This leads us to expect that, firstly, different lines of the dynamic processes will not converge at a single point corresponding to the current I_m (since the values of I_m vary with B); secondly for the same reason the maxima on the lines of the dynamic processes in the $R_{nz}(I)$ diagram will not lie on a single vertical.

The foregoing assumptions are excellently confirmed by experimental data. Figures 6-10 and 6-11 show the $R_{nz}(I)$ and $U(I)$ relationships obtained by the single-solenoid method for various values of the initial current. The first of these figures refers to

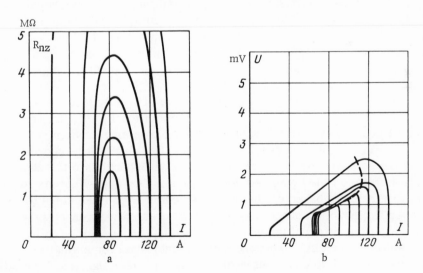

Fig. 6-10. The $R_{nz}(I)$ and $U(I)$ relationships obtained by the single-solenoid method for a seven-strand cable with an indium filler.

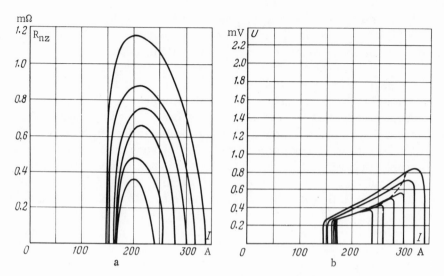

Fig. 6-11. The $R_{nz}(I)$ and $U(I)$ relationships obtained by the single-solenoid method for a nineteen-strand cable with an indium filler.

a solenoid made from a seven-strand cable (four superconducting and three copper strands) in an indium filler, the second to a nineteen-strand cable (six superconducting and thirteen copper strands) in an indium filler. The first solenoid has an internal diameter of 30 mm, an external diameter of 82 mm, a height of 77 mm, 830 turns, and an inductance of 0.0144 H; the figures for the second solenoid are 30, 72, and 77 mm, 356 turns, and 0.0023 H, respectively.

We see from the figures that the value of I_m for the seven-strand conductor is 65-68.5 A; the maxima of the R_{nz} (I) curves for different I_0 are considerably displaced from one another.

In Figs. 6-10b and 6-11b (as before) the broken lines represent the geometrical locus of the points corresponding to the equilibrium states of the combined conductor for $T \geq T_{c0}$. Whereas for the experiments based on the double solenoid each line refers to a specific induction of the magnetic field created independently with the aid of the external solenoid 2 (Fig. 6-1), the lines of equilibrium states determined from the experimental data for a single short-circuited solenoid refer to the case in which the field created by the coil varies with varying current. The line of equilibrium states recorded in such an experiment is related to the lines of equilibrium states obtained for different values of B = const in the manner illustrated schematically in Fig. 6-12. In this figure the broken lines correspond to constant values of the magnetic-field induction in which $B_1 > B_2 > B_3$. The continuous line corresponds to the varying magnetic fields encountered in the course of experiments carried out with the single short-circuited solenoid. It is clear that the greater the value of I_0, the greater will be the value of I, and hence B, at a point on the line of the dynamic process corresponding to the equilibrium state.

In principle, the line of states based on the experimentally obtained data with the single, short-circuited solenoid might also be obtained by the two-solenoid method. In this case the external magnetic field would have to be varied synchronously with the changing current of the dynamic process, in accordance with the particular solenoid characteristic B(I) which would apply if the winding were normal rather than bifilar. Such an experiment, however, would be very complicated.

It should be emphasized that the basic laws underlying the propagation of the normal zone established from the results of experiments with short-circuited coils are also applicable in the case of coils connected to an external supply source (with due allowance, of course, for the properties of that source). The greater the inductance of the short-circuited solenoid, the more slowly (other conditions being equal) will the current diminish in the dynamic process, and hence the closer will the solenoid in question be, as regards its dynamic characteristics, to a solenoid connected with a supply source for I = const.

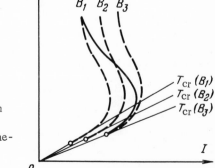

Fig. 6-12. Lines of equilibrium states determined from experiments based on the two- and one-solenoid methods.

We remember that, when using the double-solenoid method, the test coil is made in the bifilar manner, i.e., between every two turns with the same direction of the current is a turn with the current traveling in the opposite direction. As a result of this, when a normal zone arises in one of the turns of such a coil, the following situation will be created: After occupying one turn completely, the normal zone will start traveling through the following turns; thereupon the neighboring turns in the normal state will be separated by turns still in the superconducting state. In order to capture the turns causing currents in the opposite direction, the normal zone will first have to proceed to the end of the solenoid and then spread along the turns in question. Thus the turns of the coil existing in the normal state will be separated by relatively large gaps. This concept also applies to real (not bifilar) coils in which either the gap between the turns or the diameter of the winding is large. In the latter case the normal zone propagating along the turn will only reach the point in the neighboring turn lying alongside the point of generation of the normal zone after a relatively long period of time. When a normal zone propagates in the loose coil of a small-diameter single solenoid the turns existing in the normal state lie much closer to one another (neighboring turns).

In the cases under consideration any turn in the coil may experience very different conditions of heat transfer from the surface. On passing through the boiling crisis a vapor film is formed on the surface of the conductor, separating the surface from the liquid helium and possessing a considerable thermal resistance. In the case of a bifilar (or similar) coil, the surface vapor films on the neighboring normal turns are separated from one another by a fairly large gap (Fig. 6-13a). In the case of an ordinary winding of small diameter the surface films, lying close to one another, may clearly merge (Fig. 6-13b). This merging of the films on neighboring turns, equivalent to an increase in the effective film

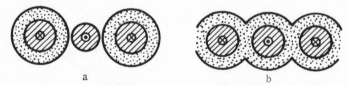

a   b

Fig. 6-13. Vapor films on the surface of a conductor for (a) a
bifilar winding and (b) an ordinary winding.

thickness, will increase the thermal resistance of the layer separating the surface of the combined conductor from the liquid helium, and hence will increase the temperature of the conductor,
which will have a considerable effect on the parameters of the normal-zone propagation process.

In order to verify this we set up a special experiment. From
an eight-strand conductor without any insulation we made a bifilar
layer, placed inside a Manganin wire coil. This layer was connected to the supply superconducting solenoid as in the experiments
based on the double-solenoid method. The Manganin coil with the
bifilar layer was placed inside another superconducting solenoid,
which was used to create the magnetic field. Thus the experimental arrangement was similar to that of Fig. 6-1.

Diagnostic microheaters (producing a normal zone in the usual way) were located on neighboring turns of the bifilar layer, i.e.,
on turns with opposite current directions. These heaters were
placed, not one under the other, but with a certain mutual displacement, so as to avoid the overheating of the conductor when
the heaters were operated simultaneously. As before, the microheater was applied for a short period of time (roughly 0.1 sec),
the power required to produce the normal zone being chosen empirically. Provision was made for the possible synchronized application of both heaters.

Potential leads were soldered to the conductor on both sides
of one of the microheaters (at a distance of 1 cm from each other).
The potential difference U_{meas} from this measuring section was
applied to the input of a two-coordinate automatic recorder; the
signal from a bismuth magnetoresistive detector placed inside the
supply solenoid was applied to the second input. Thus the recorder
traced the U_{meas} (I) relationship. At the same time the U_{nz} (I) relationship, where U_{nz} is the potential difference at the ends of the
bifilar layer under consideration, was also recorded by means of
an automatic recorder.

Before carrying out the main series of experiments we tried
two preliminary series. The U_{nz} (I) relationship was plotted for
the same values of the initial current I_0, once on introducing the
normal zone into the coil by means of one of the microheaters, the
second time by means of the other. Within the limits of experimental error of the automatic recorder employed the U_{nz} (I) lines
obtained in the two cases were identical. This indicates, firstly,

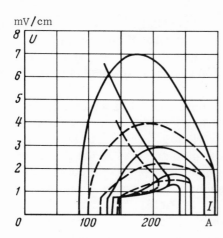

Fig. 6-14. Volt—ampere character-istics of the sample obtained by con-necting one microheater (continuous lines) and two microheaters (broken lines).

the uniformity of the combined conductor sample used in preparing the bifilar layer, secondly, the correctness of the power chosen for each of the microheaters, and thirdly, the identity of the heat-transfer conditions for both turns in question.

We conducted two main series of experiments. In the first series the U_{meas} (I) relationship was determined for the case in which only one microheater was connected — the one placed inside the measuring section. The results of these experiments are shown as continuous lines in Fig. 6-14.

The second series of experiments was carried out with both microheaters connected. In this case a normal zone arose in neighboring turns, and the vapor film developing on the surface of the conductor after passing through the boiling crisis was able to overlap as shown in Fig. 6-13b. The experiments were continued up to the same values of the initial current as in the first series, and the U_{meas} (I) relationship was recorded on the same sheet. The results of the second series of experiments are shown as broken lines in Fig. 6-14.

We see from the figure that the experimental results confirm our assumption as to the overlapping of the surface vapor films on neighboring turns of the coil — the line of equilibrium states obtained from the second series of experiments (with both the micro-heaters connected) corresponds to smaller currents than the line obtained from the first series (one microheater connected).

Figure 6-15 shows the current I* as a function of the magnetic field induction, the two curves corresponding to the first and sec-

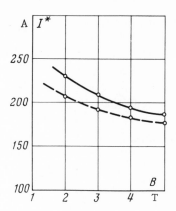

Fig. 6-15. Current I* as a function
of the magnetic field induction on
connecting one microheater (continuous line) and two microheaters
(broken line).

Fig. 6-16. Arrangement for measuring the propagation velocity of
the normal zone.

ond series of experiments (continuous and broken lines respectively). The substantial decrease in the value of I* indicates a
considerable worsening of the conditions of heat transfer in the
second series of experiments. By analogy with Eq. (6-7) it is easy
to derive the following relationship between the effective heat-
transfer coefficients for the cooling conditions in question:

$$\frac{h_2}{h_1} = \left(\frac{I^*_2}{I^*_1}\right)^2 . \qquad (6-9)$$

In the case under consideration the h_2/h_1 ratio averages 1.2.

It is extremely interesting to make an experimental determination* of the propagation velocity of the normal zone along the
combined conductor v. This velocity was measured in the following way. Potential leads a, b, and c were soldered to the test sample of the combined conductor, close to the site of the diagnostic
microheater (Fig. 6-16). The distance between the points of attachment of the leads a and b was 10 cm. The leads a and c and b

---

*This question is considered theoretically in Section 6.4.

and c were connected to an oscillograph. At the same time the
current in the test solenoid I was measured in the normal fashion.

After a nucleus of normal zone had been created in the solenoid
by means of the microheater M, this zone started traveling along
the conductor. The oscillograph ribbon recorded the instants t_1
and t_2 at which a potential difference appeared at the points a and b,
i.e., the instants at which the leading edge of the normal zone
passed points a and b. From known values of t_1 and t_2 and the distance l between points a and b we determined the propagation velocity of the normal zone to be

$$\mathfrak{v} = l/(t_1 - t_2). \tag{6-10}$$

Since the distance and hence the resistance of the section a–b
in the normal state are relatively small, when the normal zone
travels from a to b the current "frozen" in the test solenoid
changes very little. Hence the measured value of the propagation
velocity of the normal zone may to a high degree of accuracy be
referred to the average current

$$I_{av} = \frac{1}{t_1 - t_2} \int_{t_1}^{t_2} i(t)\, dt. \tag{6-11}$$

The velocity $\mathfrak{V}$ corresponding to the contraction of the normal
zone is measured in the same way [for $i(t) < I_p$].

It should be noted that the propagation velocity of the normal
zone may be calculated from the $R_{nz}(t)$ relationship, which is de-

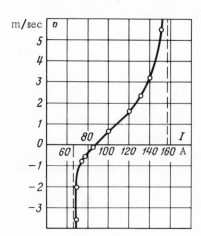

Fig. 6-17. Propagation velocity
of the normal zone as a function
of current for a seven-strand cable.

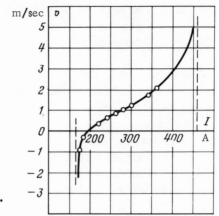

Fig. 6-18. Propagation velocity of the normal zone as a function of current for a nineteen-strand cable.

termined in the analysis of experimental data obtained by the single-solenoid method:

$$\mathfrak{V} = \frac{1}{r}\frac{dR_{nz}}{dt}, \qquad (6\text{-}12)$$

where r is the resistance of unit length of the combined conductor. However, the accuracy of velocity determination by this method is greatly inferior to the accuracy achieved by direct experiment.

The propagation velocity of the normal zone along the combined conductor in a loose coil measured as a function of current by direct experiment is illustrated in Fig. 6-17 (seven-strand cable) and Fig. 6-18 (nineteen-strand cable). These results are in excellent agreement with the earlier developed concepts of the characteristic currents I_m and I_p.

The minimum current for the existence of a normal zone I_m was determined from the resultant $\mathfrak{v}(I)$ relationship as the current for which (in the limit) an unlimited increase in the propagation velocity of the superconducting zone should occur. For this purpose the resultant data were expressed in the coordinate system $1/\mathfrak{v} = f(I)$; the curve passing through the experimental points was extrapolated until it intersected the horizontal axis, and the value of I_m was established from the point of intersection.

The minimum current for the propagation of the normal zone I_p was determined as the current for which $\mathfrak{v} = 0$, as indicated by the point of intersection of the $\mathfrak{v}(I)$ relationships with the horizontal axis.

In an analogous way, by using the normal-zone propagation velocities, we may determine the critical coil current I_c. This method of determining I_c may be of interest for large unstabilized coils.

The values of I_m and I_p found in the ways just discussed agree closely with the data obtained from the $R_{nz}(I)$ relationship (Figs. 6-10 and 6-11).

Thus these experiments once again confirm the validity of the conclusions formulated in Chapter 5 as regards the character of the volt—ampere chracteristics for the equilibrium states of a combined conductor, and reveal the limits of applicability of the foregoing model representations for describing dynamic processes in superconducting coils.

6.3. Influence of Cooling the Coil
with Superfluid Helium

A number of authors have suggested that for cooling loose superconducting coils helium cooled below the λ point (2.19 K), i.e., helium-II, should be employed. Helium-II has also been proposed for "almost" compact coils characterized by enthalpy stabilization. Superfluid helium occupies the narrow channels in the winding which resist the penetration of helium-I and increases the effective enthalpy of the coil by virtue of its own heat of vaporization. The desirability of using helium-II arose from two considerations: firstly, the fact that with decreasing temperature the critical current of the superconductor increases; secondly, the fact that the use of helium-II provides additional advantages from the point of view of the coil-cooling conditions, since the superfluid helium, which has an infinitely great thermal conductivity, passes easily into the channels between the turns.

Whereas the first of these considerations is noncontroversial, the second requires careful verification. There is little doubt that the use of helium-II for cooling the coil has indisputable advantages. However, this situation only holds until the temperature of the conductor surface exceeds the temperature of the boiling crisis;[*]

---

[*]Strictly speaking it is better not to refer to the boiling crisis, which in the case of helium-II is not really a valid concept, but to the critical thermal flux after which a vapor film is formed on the cooled surface.

a vapor film having a greater thermal resistance is then formed
on the surface. In this respect it is not known a priori what
changes the use of helium-II instead of helium-I will induce. We
accordingly set up some experiments for purposes of verification
[76].

We studied the short-circuited superconducting solenoid already described, its windings being made of nineteen-strand cable;
the helium channels between the turns of the coil were created in
the usual manner by winding a Capron filament around the cable.

At one of the inner points of the coil we attached a diagnostic
microheater to the cable. Measurements were carried out by the
single-solenoid method using the following parameters: internal
and external coil diameters 30 and 72 mm, coil height 77 mm, inductance 2.3 mH, critical current 460 A at $T_t = 4.2$ K.

The dynamic process developing after the introduction of a
normal-zone nucleus into the coil was studied at $T_t = 4.2$ and 1.9 K.
The two experiments were carried out for the same initial current
I_0 equal to 400 A. The results are presented in Fig. 6-19.

We see from Fig. 6-19a that the maximum resistance of the
normal zone R_{nz} at $T_t = 1.9$ K is about 1.5 times greater than at
$T_t = 4.2$ K. Figure 6-19b gives the values of the resistance r_{nz} of
a short (roughly 1 cm long) section of the test cable close to the
microheater. This section, as already mentioned, existed under

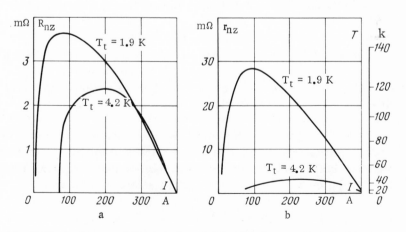

Fig. 6-19. Resistance of the normal zone (a) and a short section of cable close
to the microheater (b).

practically isothermal conditions and played the part of a resistance thermometer; the resistance r_{nz} increases with rising temperature chiefly by virtue of the corresponding rise in the resistivity of the substrate material.

For comparison Fig. 6-19b gives an approximate temperature scale calculated on the basis of the [77] data from the $\rho(T)$ relationship for copper, in which $\rho_{273K}/\rho_{4.2K} = 100$.

We see from this figure that on using helium-II the maximum temperature of the loose winding in the dynamic process amounts to about 130 K, roughly three times greater than on using helium-I.

This seemingly unexpected result may clearly be explained in the following way. When the normal zone appears, film-type boiling develops on the surface of the combined conductor. Other conditions being equal, the effective thermal conductivity of the film, which is also clearly associated with various convective mechanisms, is lower, the lower the vapor density and hence the lower the pressure. Since the pressure inside the vapor film is practically equal to the saturation pressure at the temperature in question, as the temperature of the helium tank is reduced the saturation pressure, and hence the effective thermal conductivity of the vapor film, rapidly diminish (for T = 1.9 K ρ_s = 0.0023 MPa). The reduction in the heat-transfer coefficient from the surface of the combined conductor with decreasing temperature explains the rather large rise in coil temperature which occurs when the normal zone develops. Ultimately this means that the maximum equilibrium current I* diminishes with decreasing temperature of the helium tank; usually I^*_{He-II} lies much lower than the critical current.

It follows that when the normal zone is created the short-circuited loose coil cooled with helium-II will be heated to much higher temperatures for the same initial current than a coil cooled with helium-I. For large coils this rise in temperature may be so great that the coil will burn out.

In other words, since the maximum equilibrium current I* decreases on changing to He-II cooling, the reliability of any superconducting coils connected to a supply source guaranteeing constancy of the currents, as well as large short-circuited coils,*

---

* As already mentioned (Section 6.2), large short-circuited coils, having a high inductance, are similar in their dynamic properties to coils connected with a supply source guaranteeing a constant current.

will be greatly reduced. Of course, this conclusion holds not only for cooling with helium-II but also for cooling with helium-I in the course of evacuation, since as pressure decreases the thermal resistance of the vapor film on the cooled surface becomes greater. A situation may also arise in which a transition of the loose coil into the normal state, not leading to any emergency in the case of cooling with helium-I, will lead to the burning-out of the coil in the case of cooling with helium-II. On using partly stabilized super-conducting magnetic systems the most dangerous case is that in which $I^*_{He-I} > I_{ct}$ (so that for $T = 4.2$ K the system is reliable up to $I = I_{ct}$) but $I^*_{He-II} < I_{ct}$.

Thus the idea that the use of superfluid helium (or helium-I with evacuation) always improves the operating conditions of a loose superconducting coil is incorrect. The use of helium-II is only desirable if the critical thermal flux from the surface of the combined conductor is not exceeded. Otherwise the system may only operate reliably over a range of currents not exceeding I^*_{He-II}.

6.4. Rate of Propagation of the Normal

Zone along a Combined Conductor

The processes underlying the growth (or contraction) of the regions occupied by the normal and/or resistive zones taking place in the coil of a superconducting mag-netic system largely determine the safety of the latter. Hence the velocity of the normal-zone boundary is an extremely important parameter characterizing the par-ticular conductor. The dependence of this velocity on the current may be studied most accurately for the simplest case in which the normal zone only propagates along the conductor.

For a theoretical analysis of the motion of the normal zone we must in this case consider Eq. (5-1), which describes the changes taking place in the temperature distribution of the conductor with time. Generally speaking, we should allow for the temperature dependence of the coefficients in this equation, as these may vary over extremely wide ranges. The most temperature-dependent quantity is the specific heat of the conductor material c (in the range of present interest, c is proportional to T^3) and also the thermal conductivity λ. However, a detailed analysis of an equation of type (5-1) with variable coefficients is extremely difficult for the present analysis. We shall therefore confine attention to a simplified model in which these coefficients are constant. Such an approximation is quite justified when making a qualitative study of the velocity of the normal regions. The quantitative relationships so derived may be used for practical estimates by substituting certain averaged quantities into them, these being characterisitic of the corresponding temperature ranges.

In the approximation under consideration it is again convenient to use the dimensionless form (5-2) of the original equation (5-1):

$$\frac{\partial \tau}{\partial \theta} = \frac{\partial^2 \tau}{\partial X^2} - \tau + \alpha \tilde{r} \iota^2. \tag{6-13}$$

The steady motion of the normal (or superconducting) region occurring a fair time after the onset of the transition should take place in the form of a simple displacement of the temperature profile at a certain constant dimensionless velocity c. We must thus construct solutions of Eq. (6-13) in the form

$$\tau = \tau(X - c\theta), \tag{6-14}$$

which will tend toward the values characterizing the equilibrium states for the corresponding regions as the argument tends toward $\pm \infty$. We thus have to assume that equilibrium states exist (in other words, that there exists a value of $I^* > I_{ct}$), since if this is not so [61] there may be no such quasi-stationary type of propagation at all (see Section 6.2).

Substituting the solution of form (6-14) into Eq. (6-13) we obtain

$$\frac{\partial^2 \tau}{\partial X^2} + c \frac{\partial \tau}{\partial X} - \tau + \alpha \iota^2 \tilde{r} (\tau, \iota) = 0. \tag{6-15}$$

The desired solutions should tend toward the steady value $\alpha \iota^2$ as the argument X tends toward $-\infty$, and to zero as X tends toward $+\infty$. It is easy to see that in this case a positive velocity c corresponds to motion of the normal zone from left to right, i.e., this quantity expresses the velocity of the normal zone itself. A negative value of c represents the propagation of the superconducting zone, i.e., the motion of the normal zone in the negative X direction.

As an extremely simple example let us find the $c(\iota)$ relationship for the case in which the resistance r changes sharply from zero to unity at the point $\tau = 1 - \iota$. We place the origin of coordinates at this point. It is not difficult to see that the solution of Eq. (6-15) for $X < 0$ should take the form

$$\tau = \alpha \iota^2 - (\alpha \iota^2 - 1 + \iota) \exp \left[\sqrt{\frac{c^2}{4} + 1} - \frac{c}{2} \right] X. \tag{6-16}$$

Analogously for $X > 0$ we obtain

$$\tau = (1 - \iota) \exp \left[-\frac{c}{2} - \sqrt{\frac{c^2}{4} + 1} \right] X. \tag{6-17}$$

The condition for determining c is obtained by requiring the derivative $\partial \tau / \partial X$ to be continuous at the point $X = 0$:

$$(\alpha \iota^2 - 1 + \iota) \left(\sqrt{\frac{c^2}{4} + 1} - \frac{c}{2} \right) = (1 - \iota) \left(\frac{c}{2} + \sqrt{\frac{c^2}{4} + 1} \right). \tag{6-18}$$

Hence by making some simple transformations we arrive at Eq. (5-3) which was first developed by Keilin et al. [59].

Our $\dot{r}(\tau)$ relationship is characterized by the fact that the velocity tends to $+\infty$ and $-\infty$ as the current ι tends to unity and ι_1^*, respectively, in accordance with Eq. (5-4).

This may be explained in the following way. The states of the conductor for the current values indicated in the present model are absolutely unstable, and therefore an infinitely high velocity of the transition means a practically simultaneous transition along the whole length of the conductor. The propagation of the normal zone should clearly have a character very close to this for conductors in which, owing to the presence of, for example, a substantial internal thermal resistance, the $r(\tau)$ relationship has a jump-like character (Fig. 4-10).

It is interesting to consider a certain simplification of Eq. (5-3) for cases in which cooling of the conductor is insufficiently rapid. We then have $\alpha \to \infty$ and Eq. (5-3) takes the following simple form

$$c \approx \sqrt{\alpha}\, \iota / \sqrt{1-\iota}. \qquad (6\text{-}19)$$

On passing to dimensional notation the dependence on the heat-transfer coefficient h vanishes

$$\mathfrak{v} \approx \frac{\sqrt{\rho_{sub}\gamma_{sub}}}{c\gamma} \frac{I}{\sqrt{(1 - I/I_c)\,(T_{c0} - T_t)}}, \qquad (6\text{-}20)$$

where γ is the density of the conductor material.

In this form the equation is applicable for not too small values of the current, and is often used for estimating the rate of propagation of the normal zone when calculating transitional processes in superconducting coils.

The rate of normal-zone propagation may also be analyzed in a similar way for the case in which the $\tilde{r}(\tau)$ relationship corresponds to a particular type of conductor. However, even for the only slightly more complex case of $\tilde{r}(\tau)$ in the form (5-5), it is difficult to obtain analytical expressions for the velocity, and we have to confine attention to numerical computer calculations. The results of such calculations [78] are presented in Fig. 6-20 in the form of a relationship between $c/\sqrt{\alpha}$ and the current $\iota = I/I_c$ for various values of α [in the figure the broken lines correspond to an abrupt change in r, the continuous lines to an r variation of the (5-5) type].

We see from the results presented that for large values of α the velocities, calculated for the two corresponding models in a range of currents not too close to the critical current, differ very little. On tending toward the critical current, the velocities calculated for the second model increase more slowly, so that in this limit the simplified relationship

$$c \approx \sqrt{\alpha}\, \iota \qquad (6\text{-}21)$$

is satisfied fairly closely. It should be noted that, in contrast to the first model, at the points $\iota = 1$ and $\iota = \iota_m$ the velocities tend toward certain finite limits, although the $c(\iota)$ curves approach these limiting values with a vertical tangent.

Thus the character of the transition in a long conductor close to the current values indicated differs slightly for the two models in question. Although, strictly speaking, at the extreme points the transition will take place simultaneously along the whole

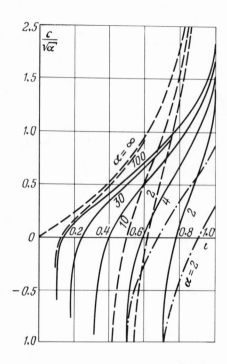

Fig. 6-20. Rate of propagation of the normal zone as a function of the dimensionless current for various values of α.

length even for the second model, since the original unstable states inevitably "decay" as a result of fluctuations, the transition at a particular point will not be related to the transitions at distant points as in the case of an infinite velocity. On the other hand, this difference in the character of the transition may to some extent be merely nominal, since, for example, the length of the resistive section will tend toward infinity at both extreme points. This means that the foregoing quasi-stationary picture of normal-zone propagation cannot be developed in any conductor of finite length close to the extreme current values.

We may here note certain characteristic quantities used in various estimates. The maximum velocity for $\iota = 1$ is given by

$$c_{max} = 2\sqrt{\alpha - 1},\qquad(6\text{-}22)$$

the minimum value for $\iota = \iota_m$ by

$$c_{min} = -2\sqrt{\alpha^{1/2} - 1}.\qquad(6\text{-}23)$$

Since $\iota_m = 1/\sqrt{\alpha}$, in Fig. 6-20 the points $c_{min}(\iota)/\sqrt{\alpha}$ lie on the arc of an ellipse

$$\frac{c_{min}}{\sqrt{\alpha}} = -2\sqrt{\iota^2 - \iota}.\qquad(6\text{-}24)$$

It should also be noted that for $\alpha \approx 1.5\text{-}20$ the derivative $\partial c / \partial \iota$ in the linear sections of the corresponding curves is almost constant and equal to approximately 6.5. For greater values of α this quantity increases, slightly exceeding the limiting value of $\sqrt{\alpha}$.

We have also calculated the rate of propagation of the normal zone for a simplified model of the boiling crisis in liquid helium (Fig. 5-6). Figure 6-20 shows only two corresponding curves (dotted and dashed lines), which are quite sufficient to reveal the slight qualitative differences arising in this case.

We see by comparison with the curves calculated without allowing for the boiling crisis that the velocity of the normal zone falls sharply, while the minimum propagation current ($c = 0$) is slightly greater. The sharp rise in velocity close to $\iota = 1$ found in the previous series of curves is almost entirely smoothed out in the new model.

An exact comparison between the results of these calculations and the experimental data for Section 6.2 is difficult in view of the fact that the values of the constants characterizing the properties of the conductor are presently not known with sufficient accuracy. However, the general form of the relationship between the rate of propagation of the normal zone and the current agrees fairly well with the first of the models considered, since as the current approaches the values $\iota = 1$ and $\iota = \iota_m$ the velocities exhibit a tendency toward unlimited growth (Fig. 6-17). This is evidently associated with the fact that the degree of stabilization of the samples studied was low, and the influence of the internal thermal resistance led to an abrupt transition. The rather sparse published data regarding the rates of propagation of the normal zone for strongly stabilized samples [79] nevertheless enable us to conclude that there is no marked rise in velocity in these samples when tending toward the critical current.

Chapter 7

Combined Conductors with Forced Cooling

7.1. General Principles

In the foregoing discussions we have considered the state of thermal equilibrium of combined conductors (with a uniform longitudinal temperature distribution, or else in the presence of a temperature gradient) for the case in which the conductor is immersed in a liquid helium tank. However, there is one special case in which the combined conductor is made hollow and the helium passes along its inner surface.

Such conductors may have a radial configuration (Fig. 7-1) and may be cooled either with liquid helium (or a vapor—liquid mixture) or else with helium in the supercritical state. Superconducting coils made from hollow conductors (coils with forced circulation) in a number of cases have indisputable advantages over ordinary superconducting systems. Firstly, such coils offer a more flexible regulation of the cooling conditions by varying the rate of flow of the liquid helium through the conductor. Moreover, considerably less helium is required for filling the system, and its reliability therefore increases from the point of view of explosion hazard should there be an accidental transition into the normal state (Section 3.1). Secondly, the amount of helium needed for the preliminary cooling of the coil to the working temperature is reduced because of the fuller use made of the cooling capacity of the departing vapor. Thirdly, the use of hollow combined conductors enables us to provide a greater electrical and mechanical strength of the coil without worsening the cooling conditions, and greatly simplifies the construction of the cryostat.

It should nevertheless be noted that the creation of large superconducting coils with forced circulation encounters certain practical difficulties. This is associated with the fact that the dimensions of such coils are limited by the hydraulic resistance of the conductor to the flow of liquid helium. The effective hydraulic resistance of the coil may be reduced by dividing it into sections and later connecting the sections in parallel. However, the parallel arrangement does not ensure an identical rate of helium flow through the various sections, especially if a normal zone forms in one of them. The development of such a zone leads to an increase in hydraulic resistance of

217

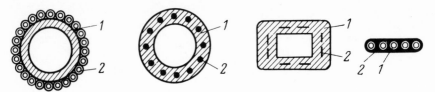

Fig. 7-1. Configuration of hollow combined conductors. 1) Copper; 2) superconductor.

a section of the conductor and ultimately to a reduction in the rate of helium flow through the section.

Other disadvantages of superconducting coils with forced circulation include the danger that the channels may be blocked by vapor locks. A possible way of solving this problem is the use of helium in the supercritical state as coolant [80]. Since the temperature corresponding to the critical point for helium (T_{cr} = 5.2 K) is only a little above the boiling point of liquid helium at atmospheric pressure (4.2 K), the use of helium in the supercritical state causes no serious reduction in the current-carrying capacity of the superconductor or in its stability.

The first attempt at using helium in the supercritical state for cooling a solenoid winding was made by Morpurgo [81]. In the experimental solenoid in question the helium was circulated by means of a pump which created a pressure drop of 0.01-0.015 MPa for a helium flow of up to 420 liters/h.

In the Kurchatov Institute of Atomic Energy another circulation system was employed; this was called a "compensatory system" and constituted a very simple liquefaction or refrigeration system [82, 83]. In this case the liquid helium passed successively through all the sections of the solenoid, which had a fairly large hydraulic resistance (up to 1.3 MPa), the helium flow rate not exceeding 20 liters/min.

In considering the advantages and disadvantages of systems with forced cooling we must mention yet another field of application of circulation systems — superconducting electrical transmission lines, for which the use of circulatory cooling systems is the only solution.

7.2. Theory of the Combined Conductor with Forced Cooling

It was assumed in Section 4.1 that the temperature T_t of the coolant was constant, i.e., that the heat capacity of the helium tank was infinitely great. For a combined conductor with forced cooling this assumption cannot of course be made, since the thermal flux q carried from the surface of the conductor into the helium is limited. In general form the flux is determined by the equation

$$q = M \int_1^2 c \, dT, \qquad (7-1)$$

where c is the specific heat of helium, M is the mass rate of helium flow, and 1 and 2 are indices corresponding to the initial and final states.

The Joule losses in the conductor $W = \rho I^2 L/A$ should not exceed the thermal flux defined by (7-1). This imposes a limitation upon the maximum length of the normal section. If the transition of the conductor from the superconducting into the normal state takes place suddenly (Fig. 5-2a), the maximum length of the normal section should not exceed a critical value given by

$$L^* = qA/\rho I^2. \tag{7-2}$$

In addition to this, if the temperature distribution along the combined conductor directly immersed in a liquid helium tank were symmetrical (Fig. 5-7), for a conductor with forced cooling the temperature profile is asymmetrical (Fig. 7-2). This is because the rates of propagation of the normal zone along and against the helium flow differ.

When the length of the normal section is less than the critical value L*, the normal zone moves along the flow, contracting in dimensions. By analogy with the earlier terminology we shall call such a conductor completely stabilized.

If the length of the normal section exceeds L*, the following cases may occur:

1. The normal zone increases in dimensions, but its upper boundary moves along the flow (the upper boundary of the normal zone, reckoned along the flow and corresponding to the transition from the superconducting to the normal state, will subsequently be called the "upper boundary," and the lower boundary corresponding to the transition from the normal to the superconducting state the "lower boundary").
2. The normal zone increases in dimensions, its upper boundary moving against the helium flow.

Since the real conductor has a finite length, in the first case, after the upper boundary of the normal zone has reached the lower end of the conductor, the normal zone vanishes. Thus such a conductor may be regarded as pseudostabilized. In a conductor of infinite length the normal zone will move along the flow, increasing in dimensions.

In the second case the normal zone will grow in both directions (along and against the flow) until the whole conductor is filled. In order to restore the superconducting state the current must be reduced. Clearly such a conductor will be unstabilized.

The problem of the propagation of the normal zone along a conductor with forced cooling was first considered in [84]. The differential equations describing the state of

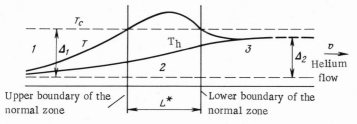

Fig. 7-2. Temperature profile of a conductor with forced cooling.

an element of the conductor take the form

$$c_l \frac{\partial T}{\partial t} = \lambda_l \frac{\partial^2 T}{\partial x^2} + \frac{\rho I^2}{A} - hP(T - T_h);$$ (7-3)

$$c_h \frac{\partial T_h}{\partial t} = hP(T - T_h) - c_h \mathfrak{v}_h \frac{\partial T_h}{\partial x},$$ (7-4)

where T and T_h are the temperature of the conductor and coolant (helium), c_l is the heat capacity of the conductor (per unit length), c_h is that of the coolant, λ_l is the thermal conductivity of the conductor (per unit length), $\mathfrak{v}_h$ is the rate of cooling flow, ρ is the resistance of the conductor in the normal state, and x is the coordinate along the conductor.

For an origin of coordinates moving at a velocity $\mathfrak{v}_1$ with respect to the conductor, Eqs. (7-3) and (7-4) transform in the following way:

$$c_l \frac{\partial T}{\partial t} = \lambda_l \frac{\partial^2 T}{\partial x^2} + c_l \mathfrak{v}_1 \frac{\partial T}{\partial x} + \frac{\rho I^2}{A} - hP(T - T_h);$$ (7-5)

$$c_h \frac{\partial T_h}{\partial t} = hP(T - T_h) - c_h(\mathfrak{v}_h - \mathfrak{v}_1) \frac{\partial T_h}{\partial x}.$$ (7-6)

We shall solve Eqs. (7-5) and (7-6) for steady-state conditions ($\partial T / \partial t = 0$ and $\partial T_h / \partial t = 0$). For simplicity we assume that the resistance of the conductor changes abruptly from 0 to ρ on passing from the superconducting to the normal state. We use a method analogous to that of Section 5.3, i.e., we find solutions for individual intervals T within which the resistance of the conductor is equal to zero or ρ, and then "match" these solutions at the boundaries of the intervals.

S e c t i o n   1 ($\rho = 0$) (Fig. 7-2). The solution for this section is

$$T = A_1 + D_1 e^{r_2 x},$$ (7-7)

where

$$A_1 = T_h|_{x=\infty}; \quad D_1 = \Delta_1 = T_c - T_{h1}.$$

S e c t i o n   2 ($\rho \neq 0$). In this section

$$T = A_2 + B_2 x + C_2 e^{r_1 x} + D_2 e^{r_2 x},$$ (7-8)

where

$$A_2 = T_h|_{x=\infty} + \frac{1}{1 - e^{r_1 L^*}}(\Delta_1 - \Delta_2) + \frac{1}{1 - e^{-r_2 L^*}}\Delta_1;$$

$$B_2 = \frac{I^2 \rho}{A c_h(\mathfrak{v}_h - \mathfrak{v}_1) - c_l \mathfrak{v}_1} = \frac{\Delta_2}{L^*};$$

$$C_2 = \frac{\Delta_1 - \Delta_2}{1 - e^{r_1 L^*}};$$

$$D_2 = -\Delta_1 \frac{e^{-r_2 L^*}}{1 - e^{-r_2 L^*}}; \quad \Delta_2 = T_{h2} - T_{h1}.$$

(7-8a)

Section 3 ($\rho = 0$). The solution for this section may be written

$$T = A_3 + C_3 e^{r_1 x},$$ (7-9)

where

$$A_3 = T_h\big|_{x=\infty} + \Delta_2;$$ (7-10)

$$C_3 = \Delta_1 - \Delta_2.$$ (7-11)

In these relationships,

$$
\begin{aligned}
r_1 &= -\frac{1}{2}\left[\frac{hP}{c_h(\vartheta_h - \vartheta_1)} + \frac{c_1\vartheta_1}{\lambda}\right] - \left\{\frac{1}{2}\left[\frac{hP}{c_h(\vartheta_h - \vartheta_1)} + \right.\right. \\
&\quad \left.\left. + \frac{c_1\vartheta_1}{\lambda}\right]^2 + \frac{hP}{\lambda}\left[1 - \frac{c_1\vartheta_1}{c_h(\vartheta_h - \vartheta_1)}\right]\right\}^{\frac{1}{2}}; \\
r_2 &= -\frac{1}{2}\left[\frac{hP}{c_h(\vartheta_h - \vartheta_1)} + \frac{c_1\vartheta_1}{\lambda_l}\right] + \left\{\frac{1}{2}\left[\frac{hP}{c_h(\vartheta_h - \vartheta_1)} + \right.\right. \\
&\quad \left.\left. + \frac{c_1\vartheta_1}{\lambda_l}\right]^2 + \frac{hP}{\lambda_l}\left[1 - \frac{c_1\vartheta_1}{c_h(\vartheta_h - \vartheta_1)}\right]\right\}^{\frac{1}{2}}; \\
\Delta_2 &= \Delta_1 \frac{r_2 L^*}{1 - e^{-r_2 L^*}} \frac{(1 - e^{-r_2 L^*})/r_2 L^* + (1 - e^{r_1 L^*})/r_1 L^*}{1 + (1 - e^{r_1 L^*})/r_1 L^*}.
\end{aligned}
$$ (7-12)

Since at the boundary between sections 1 and 2 the second derivative exhibits a jump

$$\lambda_l \left(\frac{\partial^2 T}{\partial x^2}\right)_1 = \lambda_l \left(\frac{\partial^2 T}{\partial x^2}\right)_2 + I^2 \frac{\rho}{A},$$ (7-13)

we obtain

$$\Delta_1 \left(\frac{r_2^2}{1 - e^{-r_2 L^*}} + \frac{r_1^2}{1 - e^{r_1 L^*}}\right) - \Delta_2 \frac{r_1^2}{1 - e^{r_2 L^*}} = \frac{I^2\rho}{\lambda_l A}.$$ (7-14)

From the condition that the temperature profiles should join smoothly at the boundary between sections 1 and 2 (equal temperature gradients) we have

$$\Delta_1 \left(\frac{r_2}{1 - e^{-r_2 L^*}} + \frac{r_1}{1 - e^{r_1 L^*}}\right) - \Delta_2 \left(\frac{r_1}{1 - e^{r_1 L}} - \frac{1}{L^*}\right) = 0.$$ (7-15)

Multiplying (7-15) by r_1 and allowing for (7-14) we obtain

$$\Delta_1 \frac{r_2}{1 - e^{-r_2 L^*}}(r_2 - r_1) + \Delta_1 \frac{r_1}{L^*} = \frac{I^2\rho}{\lambda_l A}.$$ (7-16)

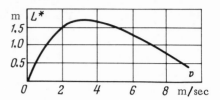

Fig. 7-3. Critical length of the normal section as a function of its velocity along the conductor.

Allowing for the condition of heat balance

$$\frac{I^2\rho}{A} = \frac{\Delta_2}{L^*}[c_h(\mathfrak{v}_h - \mathfrak{v}_1) - c_l\mathfrak{v}_1],\qquad(7\text{-}17)$$

Eq. (7-16) may be written in the following form:

$$\frac{1 + (1 - e^{r_1 L^*})/r_1 L^*}{(1 - e^{-r_2 L^*})/r_2 L^* + (1 - e^{r_1 L^*})/r_1 L^*} = -\left[\frac{c_h(\mathfrak{v}_h - \mathfrak{v}_1)}{hP}r_2 + 1\right]\frac{r_1}{r_2 - r_1}.\qquad(7\text{-}18)$$

This equation enables us to find the relationship between the velocity of the normal zone along the conductor $\mathfrak{v}_1$ and the critical length of the normal section L^*. The velocity of the normal zone is determined by the Joule losses in the conductor, the initial temperature of the coolant, and the heat-transfer coefficient from the surface of the conductor to the cooling gas. For any quantities $\mathfrak{v}_1$ and L^* connected with one another the corresponding complex $I^2\rho/AhP\Delta_1$ may be determined by means of the following relation:

$$\frac{I^2\rho}{AhP\Delta_1} = \frac{r_2}{1 - e^{-r_2 L^*}} \times \frac{[c_h(\mathfrak{v}_h - \mathfrak{v}_1)/hP][1 - c_l\mathfrak{v}_1/c_h(\mathfrak{v}_h - \mathfrak{v}_1)]}{[-r_1/(r_2 - r_1)]\{[r_2 c_h(\mathfrak{v}_h - \mathfrak{v}_1)/hP] + 1\}}.\qquad(7\text{-}19)$$

By way of example, Fig. 7-3 shows the dependence of the critical length of the normal section L^* on the velocity $\mathfrak{v}_1$, calculated for a Nb–Ti alloy wire with the following parameters (Fig. 7-4): $c_l = 0.00484$ J/cm-K, $\lambda_l = 2.102$ W-cm/K, $c_h = 0.812$ J/h-K, $hP = 0.250$ W/cm-K, $\Delta_1 = 1$ K, $\mathfrak{v}_1 = 10$ m/sec; the ratio of the copper to the superconducting cross section is 4:1.

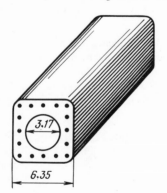

Fig. 7-4. Combined conductor with copper/superconductor cross-sectional ratio 4 : 1.

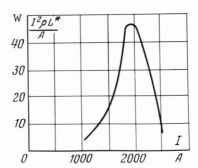

Fig. 7-5. Heat-evolution intensity
in the normal section of length L*
as a function of current.

For this particular conductor Fig. 7-5 shows the heat-evolution intensity in the
normal section of critical length L* as a function of the current in the conductor.
As the calculations carried out for this conductor show, the critical length L* reaches
a maximum for I = 1850 A, while the maximum heat evolution in the normal section
of critical length appears at I ≈ 2000 A.

It should be remembered that if the critical length L* is exceeded the normal
zone will increase. The velocities of the upper and lower boundaries of the normal
section will then differ. If other parameters are fixed these velocities depend on the
length of the normal section, and the determination of the corresponding relationships
therefore presents considerable difficulties. Fairly simple expressions may be obtained for a normal section of infinite length, since the velocity of the upper boundary
of the normal zone then asymptotically approaches a certain threshold value $\mathfrak{v}_R$ (minimum restoration velocity) and that of the lower boundary also approaches a threshold
value $\mathfrak{v}_D$ (maximum velocity of the lower boundary).

An expression may easily be found for $\mathfrak{v}_R$ from the condition of having equal
first derivatives at the boundary between sections 1 and 2 (Fig. 7-2)

$$\Delta_1 r_2 = B_2 + C_2 r_1 \tag{7-20}$$

and a jump in the second derivative

$$\Delta_1 r_2^2 = C_2 r_1^2 + \frac{I^2 \rho}{A \lambda_l}. \tag{7-21}$$

Multiplying (7-20) by r_1 and allowing for (7-21) we have

$$\Delta_1 r_2 \,(r_2 - r_1) = \frac{I^2 \rho}{A \lambda_l} - B_2 r_1. \tag{7-22}$$

Using Eq. (7-8a) we have

$$\frac{I^2 \rho}{AhP\Delta_1} = r_2 \frac{[c_h\,(\mathfrak{v}_h - \mathfrak{v}_R)/hP]\,[1 - (c_i \mathfrak{v}_R/c_h\,(\mathfrak{v}_h - \mathfrak{v}_R)]}{[-r_1/(r_2 - r_1)]\,\{[r_2 c_h\,(\mathfrak{v}_h - \mathfrak{v}_R)/hP] + 1\}}. \tag{7-23}$$

In an analogous way we may obtain the following relationships determining the
maximum velocity of the lower boundary of the normal zone $\mathfrak{v}_D$:

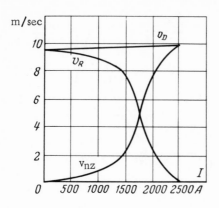

Fig. 7-6. Minimum restoration velocity v_R, maximum velocity of the lower boundary of the normal zone v_D, and rate of growth of the normal zone v_{nz} as functions of current.

for the case $\mathfrak{v}_D > \mathfrak{v}_h$ we have

$$\frac{I^2\rho}{AhP\Delta_1} = -r_1 \frac{[c_h(\mathfrak{v}_h - \mathfrak{v}_D)/hP][1 - (c_i\mathfrak{v}_D)/c_h(\mathfrak{v}_h - \mathfrak{v}_D)]}{[r_2/(r_1 - r_2)]\{[r_1 c_h(\mathfrak{v}_h - \mathfrak{v}_D)/hP] + 1\}}; \qquad (7\text{-}24)$$

for the case $\left(\dfrac{c_h}{c_h + c_i}\right)\mathfrak{v}_h < \mathfrak{v}_D < \mathfrak{v}_h$ we have

$$\frac{I^2\rho}{AhP\Delta_1} = \frac{[c_h(\mathfrak{v}_h + \mathfrak{v}_D)/hP][1 - (c_i\mathfrak{v}_D)/c_h(\mathfrak{v}_h - \mathfrak{v}_D)]}{[(r_1 + r_h)/r_1 r_2] + [c_h(\mathfrak{v}_h - \mathfrak{v}_D)/hP]}; \qquad (7\text{-}24)$$

for the case $\mathfrak{v}_D = \mathfrak{v}_h$ we have

$$\frac{I^2\rho}{AhP\Delta_1} = \frac{c^2\mathfrak{v}_D}{hP\lambda_1}. \qquad (7\text{-}26)$$

Thus on exceeding the critical length of the normal section L^* the normal zone will grow at a certain rate $\mathfrak{v}_{nz}$, which will nevertheless not exceed the difference $(\mathfrak{v}_D - \mathfrak{v}_R)$. The velocities $\mathfrak{v}_R$, $\mathfrak{v}_D$ and $\mathfrak{v}_{nz}$ calculated for the wire under consideration are shown as functions of current in Fig. 7-6. Clearly for the finite-length conductor which we have just been considering, the operating conditions at 0-2500 A are "pseudostable," since in this region the minimum restoration velocity is positive, i.e., the upper boundary of the normal zone moves along the flow and when the upper boundary reaches the lower end of the conductor its superconducting properties are completely restored.

For currents in the conductor of over 2500 A the minimum restoration velocity $\mathfrak{v}_R$ becomes negative, i.e., the upper boundary of the normal zone moves against the flow, and hence this operating condition is unstable.

7.3. Conditions of Thermal Equilibrium of the Normal Zone

As indicated in Section 5.2, in actual superconducting constructions constantly acting heat sources may exist (breaks in the superconducting strands, solder joints, nonideal superconducting compounds, etc.), and in principle these may create a normal zone. These additional sources of heat evolution may be taken into account in the theoretical model in a way similar to that discussed in Section 5.2.

Let us place a point source of heat evolution of power W at the origin of coordinates. Then the equation of total heat balance for the combined conductor with forced cooling may be written in the following way;

$$c_h M \Delta_2 = \frac{I^2 \rho L}{A} + W. \tag{7-27}$$

In the steady-state condition the heat-conduction equation (7-3) for an element of length of the conductor takes the form

$$\lambda A \frac{\partial^2 T}{\partial x^2} - hP (T - T_h) + \frac{\rho I^2}{A} = 0. \tag{7-28}$$

The relationship between the temperatures of the conductor and coolant T and T_h may be found from Eq. (7-4), allowing for the fact that the mass flow rate M is proportional to the coolant flow velocity $\mathfrak{v}$

$$c_h M (T_h - T_{h^1}) = \int_0^x hP (T - T_h) \, dx. \tag{7-29}$$

Using Eq. (7-29), the heat-conduction equation (7-28) may be written in dimensionless form [85]:

$$\frac{\partial^3 \tau}{\partial x_1^3} + \beta \frac{\partial^2 \tau}{\partial x_1^2} - \frac{\partial \tau}{\partial x_1} + \alpha \beta \iota^2 r (\tau) = 0, \tag{7-30}$$

where $x_1 = x(hP/\lambda A)^{1/2}$ is the dimensionless coordinate measured along the conductor; $\tau = (T_h - T_{h1})/\Delta_1$ is the dimensionless temperature; $\iota = I/I_c$ is the dimensionless current; $r(\tau)$ is the dimensionless resistance of the conductor (for $T < T_{c0}$, $r = 0$, for $T > T_{c0}$, $r = 1$); $\alpha = \rho I_{ct}^2 / AhP(T_{c0} - T_{h1})$ is the Stekly parameter; $\beta = (h\lambda PA)^{1/2}/c_h M$.

The most important characteristic of Eq. (7-30) is the existence of a new dimensionless parameter β; the authors of the model under consideration proposed calling this the flux parameter. By solving the resultant equation, using a method similar to that of Section 5.3, and then "matching" the solution at the boundaries of the corresponding temperature ranges, we may obtain the dimensionless length of the normal zone $l = (hP/\lambda A)^{1/2} L$ as a function of the current in the conductor for various fixed values of the additional heat-evolution intensity f (Fig. 7-7).

The curves here presented give a clear picture of the behavior of the normal zone over the whole range of working currents. In Fig. 7-7 we notice the characteristic currents ι_{1cyl} and ι_{2cyl} analogous to the currents ι_1^* and ι_2^* for the model of the conductor (with a longitudinal temperature gradient) immersed in a liquid helium tank. The current ι_{1cyl} is the limiting current, on exceeding which the $l(\iota)$ curves developed unstable sections ($dl/d\iota < 0$), and corresponds to the minimum current

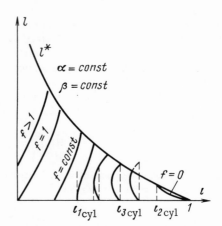

Fig. 7-7. Length of the normal
zone as a function of the current
in the conductor.

required for the existence of a normal zone. The current ι_{2cyl} is the limiting value
below which there are no stationary solutions to Eq. (7-30), and corresponds to the
minimum current for the propagation of the normal zone ι_2^*. In the particular case
$\beta = 0$ the values of the currents ι_{1cyl} and ι_{2cyl} coincide with the expressions for ι_1^*
and ι_2^* in the model involving an abrupt transition (see Section 5.2).

In Fig. 7-7 we may distinguish yet another characteristic current ι_{3cyl} which
corresponds to the case in which the normal zone, once having been created, propagates
through a distance equal to the critical length:

$$l^* = \sqrt{\frac{hP}{\lambda A}} \; L^* = \frac{1 - \iota - 2f\beta}{\alpha \beta \iota^2}.$$

Under conditions in which ι = const, on working in the current range $\iota_{3cyl} - 1$
the normal zone occupies a section of conductor longer than the critical length, so
that the normal zone will propagate along the conductor in an uncontrollable manner.

The model under consideration corresponds to the case in which the resistance
of the conductor changes abruptly on passing from the superconducting into the normal
state. Allowance for the transitional region between the superconducting and normal
zones in the present model is quite a complicated problem. However, the solution
is highly desirable, as it may greatly improve our ideas as to the conditions of stabilizing combined conductors with forced cooling, such as occurred, for example,
when studying the model of a combined conductor with a longitudinal temperature
gradient.

Chapter 8

Equilibrium and Propagation of the Normal Zone in a Close-Packed Superconducting Coil

8.1 Principal Characteristics of Close-Packed Superconducting Coils

As already mentioned there are two limiting conditions for the cooling of a combined conductor situated in a liquid helium tank. In the first of these the conductor is surrounded by a large volume of liquid helium, and in the second the conductor is wound fairly tightly and it is difficult for the helium to penetrate into the inner layers. An intermediate position between these two limiting cases is occupied by various kinds of loose windings containing channels between the turns for the passage of liquid helium.

At the present time a series of superconducting magnetic systems partially stabilized in the thermal respect are being made with close-packed windings. Chief among these are coils in which an attempt is made to obtain the maximum constructive current density and on which specific size limitations are imposed. Windings in which the method of internal stabilization is employed are also made close-packed (see Chapter 9).

A dense coil is obtained by laying the turns without any gaps. In some cases in order to increase the effective enthalpy of the coil (and sometimes the thermal conductivity as well), as well as preventing the possibility of the turns moving under the influence of the magnetic forces, the winding is impregnated with a special compound (impregnator) possessing a fairly large heat capacity

227

(or thermal conductivity), depending on the method of stabilization employed. In this way the turns are rigidly kept in place.

It was shown earlier that the case in which the combined conductor was placed in a large volume of liquid helium and the case in which a loose coil was employed were characterized by qualitatively identical laws of equilibrium and propagation of the normal zone. On using a compact coil the conditions of equilibrium of the normal zone and the laws governing its propagation differ considerably. This occurs for the following reasons.

In the case of a loose coil (and of course all the more in that of a conductor placed in a large volume of liquid helium) the turns of the combined conductor are separated by a helium layer of finite thickness. Thus the Joule heat evolving in the normal zone is passed from the surface of the conductor to the helium tank and also propagates by heat conduction along the combined conductor. Hence the propagation of the normal zone in the coil has a one-dimensional character — the normal zone only propagates along the combined conductor. An exception is the case in which the channels between the turns are fairly narrow. On passing to the film type of boiling the vapor films on the surfaces of neighboring turns may then merge with one another, and in principle the normal zone may travel not only along the conductor but also from turn to turn.

The situation is very different in a close-packed coil when the turns touch one another or are thermally connected via an impregnator. In this case the Joule heat evolving in the normal zone is propagated by heat conduction only, both along the turn in question and also to adjacent turns. Thus the propagation of the normal zone in the coil entails a three-dimensional character.

It should be emphasized that, whereas in a loose coil the expanding or contracting normal zone is continuous along the length of the combined conductor, in a close-packed coil this zone is discontinuous — sections of different turns all existing in the normal state occur close to the point at which the original normal zone was initiated. Thus in a close-packed coil normal and superconducting sections may alternate along the length of the combined conductor.

The equilibrium conditions for the normal zone in a close-packed superconducting coil were studied theoretically by Stekly [86]. The theory was based on a heat-conduction equation of the following type:

$$\frac{\partial T}{\partial t} = a_x \frac{\partial^2 T}{\partial x^2} + a_y \frac{\partial^2 T}{\partial y^2} + a_z \frac{\partial^2 T}{\partial z^2} + \frac{\beta I^2 R}{c\gamma A_\Sigma}, \qquad (8\text{-}1)$$

where a_x, a_y, and a_z are the effective thermal diffusivities along
the three coordinate axes; β is the occupation factor of the coil; the
remaining notation is the same as in Eq. (5-1), which describes
the propagation of heat along a combined conductor situated in a
large volume of liquid helium. The last term on the right-hand
side of (8-1) allows for the Joule heat evolution in that part of the
coil which exists in the normal state.

Equation (8-1) differs from (5-1) in allowing for the three-
dimensional character of the propagation of heat in the close-
packed coil and also in that there is no term acounting for heat
transfer from the surface of the combined conductor to the helium
tank. For the case in which, in addition to the Joule heat evolution
in the section of coil existing in the normal state, an internal
source of constant heat evolution also exists (e.g., a diagnostic
heater of finite dimensions, a source of heat evolution attributable
to contact resistances lying within the coil of an experimental
solenoid, and so on), Eq. (8-1) is written in the following way:

$$\frac{\partial T}{\partial t} = a_x \frac{\partial^2 T}{\partial x^2} + a_y \frac{\partial^2 T}{\partial y^2} + a_z \frac{\partial^2 T}{\partial z^2} + \frac{\beta I^2 R}{c\gamma A_\Sigma} + \frac{w_V}{c_p\gamma}, \qquad (8\text{-}2)$$

where w_V is the volume density of heat evolution from the internal
source.

Stekly [86] assumed to a first approximation that the temperature front (leading edge) propagating in a close-packed coil was
shaped like the surface of an ellipsoid. Equations (8-1) and (8-2)

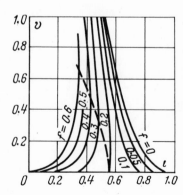

Fig. 8-1. Volt−ampere characteristics
of a closed-packed winding obtained by
calculation; v = voltage drop in the section of close-packed coil in the normal
state; f = dimensionless intensity of the
heat evolution of a point source situated
inside the coil.

were solved for the steady-state condition $(\partial T/\partial t = 0)$ in which the normal zone was in thermal equilibrium with the ambient. Solutions were obtained for the case in which the resistance of the super-conductor changed abruptly from zero to unity on reaching the critical temperature corresponding to the particular current, $T_{cr}(I, B)$. Figure 8-1 shows some volt–ampere characteristics constructed on the basis of these solutions.

As before, there are sections in which $(\partial U/\partial I)_f > 0$ which characteristically correspond to stable states of thermal equilibrium while there are sections in which $(\partial U/\partial I)_f < 0$ which correspond to unstable thermal-equilibrium states of the normal zone. The line separating the regions of stable and unstable states is shown in broken form; this line is determined by the obvious condition $(\partial U/\partial I)_f = \infty$.

We see from Fig. 8-1 that for a close-packed coil there is a characteristic current I_p^{cp}; for $I > I_p^{cp}$ all the lines $f = $ const start from the horizontal axis with a negative slope, i.e., $(\partial U/\partial I)_{f,\,U=0} < 0$, and hence in this case there are no stable states of the normal zone in the close-packed coil for any values of f.

The model of the close-packed coil under consideration is based on the assumption that the temperature dependence of the dimensionless effective resistance of the combined conductor for $\iota = $ const takes the form

$$\tilde{\rho} = \begin{cases} 0 & \text{for } \tau < 1 - \iota; \\ 1 & \text{for } \tau > 1 - \iota. \end{cases} \tag{8-3}$$

In other words no allowance is made for the transition of the

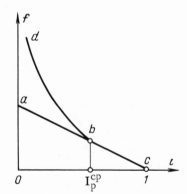

Fig. 8-2. Limits to the stability regions of a close-packed coil.

superconductor into the resistive state for $1 > \tau > 1 - \iota$ preceding the transition into the normal state.

In understanding the process taking place in a close-packed coil when a normal zone appears, an important part is played by the $f(\iota)$ [or in dimensional coordinates W(I)] relationship, illustrated schematically in Fig. 8-2 [86]. In this figure the region under the straight line $a - b - c$ corresponds to those states in which there is no normal zone in the coil, despite the existence of finite heat evolution from an internal heat source; the power of the latter is insufficient to heat the section of conductor containing the source to the critical temperature.

The region $a - b - d$ corresponds to the existence of a normal zone in stable thermal equilibrium with the ambient. If while keeping I = const we increase the power of the internal source of evolution, then in this region, corresponding to the region below the broken line in Fig. 8-1, the dimensions of the normal zone in thermal equilibrium will increase. With increasing current in the coil the range of stable thermal equilibrium of the normal and superconducting zones will diminish, and for $I = I_p^{cp}$ will vanish completely.

Finally, the region above the line $d - b - c$ corresponds to unstable states of the normal zone in the close-packed coil. This means that the heat evolution from the internal source and the Joule heat evolution in that part of the winding which has passed into the normal state are so large that for I = const the normal zone travels along the coil in an uncontrollable manner. If we increase the power of the internal source of heat evolution while keeping I = const, then on exceeding the values of W corresponding to the line $d - b - c$ the normal zone will rapidly propagate over the entire coil. As $I \rightarrow 0$ the line $b - d$ will pass to infinity; actually when I = 0 the dimensions of the normal zone will be determined simply by the power of the internal source of heat evolution, since in this case there will be no Joule heat evolution in that part of the coil which has passed into the normal state. The greater the value of f, the larger will the normal zone be; the normal zone will only fill an infinitely large coil as $f \rightarrow \infty$.

We see from the foregoing that in the current range $I < I_p^{cp}$ the appearance of the normal zone in the close-packed coil may by itself not lead to an uncontrolled transition of the whole coil into the normal state; this process will only develop in the range of

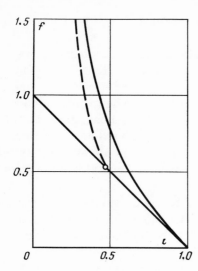

Fig. 8-3. Change in the boundaries of the stability on allowing for the division of the current between the super-conductor and the normal metal.

states above the $d - b$ line.*   In the region $I > I_p^{cp}$, when a normal-zone nucleus appears the entire coil immediately passes into the normal state.

In order to study the effect of current division in the resistive range $1 - \iota < \tau < 1$ we determined the relationship $f(\iota)$, using (5-5) to obtain the effective resistance of the combined conductor. The calculations were carried out in an electronic computer, since in this case simple analytical relationships could not be established. The volt−ampere characteristics for the new model have certain qualitative differences by comparison with the characteristics of the Stekly model. The line bounding the region of stable states in the $f(\iota)$ diagram passes through the point $\iota = 1$. Thus this region expands considerably, since the total heat evolution in this case is smaller. The boundaries of the stability region for the Stekly model (broken lines) and for the model here considered (continuous lines) are illustrated in Fig. 8-3.

A more detailed investigation into this problem is pointless in view of the severe simplifications introduced in the model analysis of the heater. It should also be remembered that in loose coils it is a more frequent practice to use combined conductors with an insufficient degree of stabilization. In such conductors the internal thermal resistance of the superconductor may play a considerable part and the $r(\tau)$ relationship may assume a more complicated form (Fig. 4-10).

---

*We remember that in a loose coil for $I < I_p$ (or more strictly $I < I_m$) the total range of normal states is stable. The greater the value of W for a given I = const, the greater is the size of the normal zone in the loose coil, but for any size this zone still remains stable.

It is interesting to compare the $W_h(I)$ relationships for the loose and close-packed coils. For a certain range of currents in the loose coil the normal zone is in stable thermal equilibrium for any values of W_h; the length of the section of conductor occupied by the normal zone then increases in proportion to rising W_h. In the close-packed coil, for any value of the current (except $I = 0$), when the power of the heater exceeds a certain value an avalanche process extends the normal zone throughout the entire coil. This may be explained in the following way. The greater the value of W_h, the greater will be the section of coil occupied by the normal zone, and hence the greater the intensity of Joule heat evolution in this section. However, in a loose coil, in which the propagation of the normal zone has a one-dimensional character, the specific thermal flux from the surface of the combined conductor to the helium tank due to Joule heat evolution in the normal section does not alter on changing the length of the normal section. If in fact l is the length of the section occupied by the normal zone, the area of the cooled surface of this section will be $S = Pl$, where P is the cooled perimeter of the conductor, while the Joule heat evolution in this section will be $Q_J = I^2 \rho l / A$, where A is the cross-sectional area of the conductor. It is clear from this that the specific thermal flux q_J, equal to Q_J / S, is independent of l.

Thus on increasing W_h the boundary of the normal zone moves forward, while by virtue of heat conduction along the conductor the thermal flux remains constant.

In a close-packed coil in which the propagation of the normal zone has a three-dimensional character, as the dimensions of this zone increase (in conformity with increasing W) the volume occupied by the zone, which generates Joule heat, increases faster than the surface of the normal zone through which this heat is carried away. Thus with increasing W the specific thermal flux through the surface of the normal zone increases, and on reaching a certain intensity it exceeds the critical value. The equilibrium between the generated and outflowing heat is then broken, the temperature of the coil rises, and the normal zone propagates along the close-packed winding in an uncontrollable fashion.

8.2. Experimental Results

We carried out an experimental investigation into the equilibrium conditions of the normal zone in a close-packed winding in the same manner as in Chapter 5, in which we studied the equilibrium conditions for a loose coil. This method is based on using an adjustable low-resistance shunt connected to the test solenoid (having a close-packed bifilar winding), a diagnostic heater being placed at one of its internal points. The shunt and test solenoid were connected to a supply source.

For a specified constant power of the microheater we recorded the voltage drop in the coil as a function of coil current with the help of a two-coordinate automatic recorder. At the same time, using a second two-coordinate automatic recorder, we recorded the voltage drop U_{meas} at the ends of a short (0.5 cm) auxiliary measuring section with a microheater inside it. As before, the use of the low-resistance shunt enabled us to realize the unstable parts of the volt−ampere characteristics for W = const.

Figures 8-4 and 8-5 present the experimental data obtained for close-packed coils respectively made from eight- and four-strand conductor at B = 2 T. The broken lines represent the parts of the volt−ampere characteristics for W_h = const not realized in the experiments. In the range of states corresponding to these sections, the nonattenuating current and voltage oscillations shown in Fig. 8-4 (elongated ellipse) were generated in the solenoid−shunt system connected to a supply source. These oscillations were evidently due to the fact that the coil was not ideally bifilar,

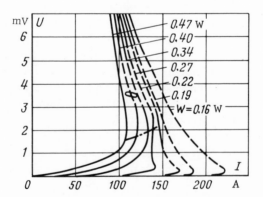

Fig. 8-4. Volt−ampere characteristics of a close-packed coil made from an eight-strand cable at B = 2 T.

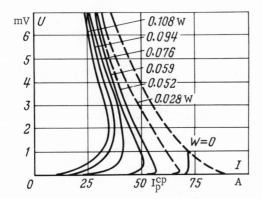

Fig. 8-5. Volt−ampere characteristics of a close-
packed coil made from a four-strand cable at B =
2 T.

so that the test solenoid might have had a certain inductance. As
a result of this the current in the test coil lags slightly and does
not correspond to the values which would be found in a circuit with
a purely active impedance. Hence the heat evolution in the winding
also lags with respect to the equilibrium values, and under certain
conditions this may lead to the development of an unstable oscillatory state. The frequency of these oscillations is fairly low, since
the processes which decide it are of the fairly slow, heat-transfer
type.

Despite the existence of these oscillations, we were able to
restore the tone character of the volt−ampere characteristics in
these regions by considering that these curves passed through the
centers of the elongated ellipses sketched by the recording pen of
the automatic recorder when the oscillatory state developed.

The volt−ampere characteristics of superconducting solenoids
with close windings illustrated in Figs. 8-4 and 8-5 agree qualitatively with the theoretical results based on the models assumed
(Fig. 8-1). At the same time there are certain differences prob-
ably associated with the fact that the form of distributed heater
assumed in these models is too simplified.

Figure 8-6 shows the volt−ampere characteristics taken from
the auxiliary measuring section in the experiments with the solenoid constructed with an eight-strand cable. The meaning of these
characteristics was discussed in full detail earlier (Section 5.6).

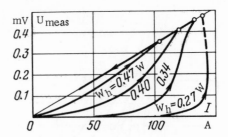

Fig. 8-6. Volt—ampere characteristics of the auxiliary measuring section obtained when studying a solenoid made from an eight-strand cable.

It follows directly from the data presented in the figure that, for each value of W_h = const, within the range of currents extending from I = 0 to a certain specific value the superconductor in the close-packed winding is in a resistive state. We see from the figure that the range of currents in which the resistive state occurs is fairly wide, and its neglect may lead to serious errors. The line T_{c0} = const constructed from the experimental data of Fig. 8-6, separating the range of resistive states from the range of normal states, is shown as a dot—dash line in Fig. 8-4.

Using a method analogous to that described earlier (Section 5.6), we carried out a series of experiments to determine the $W_h(I)$ relationship for a solenoid with a compact winding of four-strand cable. For a fixed value of the solenoid current, as the power of the microheater increased we simultaneously measured the voltage drops in the auxiliary measuring section U_{meas} and the entire solenoid U. We determined the power for which a potential difference of $\Delta U = 1\ \mu V$ appeared in the auxiliary measuring section (this power corresponds to the line $a-b$ in Fig. 8-2). We also measured the power for which the potential differences at the ends of the coil U started rising for a constant value of W_h, which indicated a transition into the region corresponding to the uncontrollable propagation of the normal zone (line $d-b-c$ in Fig. 8-2). The results of these experiments with the coil made from a four-strand cable are shown in Fig. 8-7a. We see from this figure that for this solenoid $I_p^{cp} = 56.5$ A (for $I_{ct} = 90$ A).

Figure 8-7b presents an analogous relationship obtained for a solenoid with a close-packed coil made from seven-strand indium-

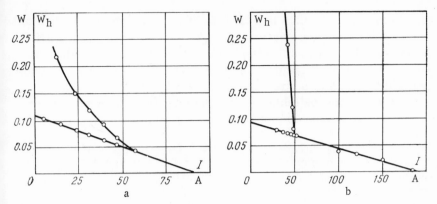

Fig. 8-7. Experimental relationships between the intensity of heat evolution from the source inside the coil and the current.

filled cable (six superconducting and one copper strand). The internal diameter of the coil is 30 mm, the external diameter 81 mm, the height 77 mm, the critical current 187 A, the number of turns 2250, the winding normal (not bifilar), and $I_p^{cp} = 49$ A.

The W_h (I) relationship shown in Fig. 8-7 is in good qualitative agreement with the results obtained for the Stekly model. The experiments so carried out, which are quite simple as regards execution technique, enable us to make a very clear determination of the current I_p^{cp} for the close-packed winding in question, and also to find the upper limit to the region of stable states for $I < I_p^{cp}$.

In order to estimate the reliability of the resultant experimental data it is especially interesting to compare the results of independent measurements relating to a close-packed winding of four-strand cable (Figs. 8-5 and 8-7a). The results of such a comparison are presented in Table 8-1; for values of W_h correspond-

TABLE 8-1

W_h, W	I_{min}, A		I_{max}, A	
	from data referring to U(I)	from data referring to $W_h(I)$	from data referring to U(I)	from data referring to $W_h(I)$
0.108	0	2.5	35	34
0.094	13	13	37	37,5
0.076	22	29	42	43
0.059	44	43	51.5	50
0.052	52	48.5	58	52.5
0.028	66	68.5	68.5	68.5

ing to the U_{meas} (I) lines in Fig. 8-6, this table gives the values of I_{min}, i.e., the current corresponding to the boundary of the region of superconducting states (line $a - b - c$ in Fig. 8-2), and also the values of I_{max}, i.e., the current corresponding to the upper boundary of the region of stable states (line $d - b - c$ in Fig. 8-2).

We see from the table that the data secured in these independent experiments agree closely with each other.

There is a slight discrepancy in the values of I_{min} and I_{max} for $W_h = 0.052$ W; this is evidently due to error in the measurement of W_h during the experiments illustrated in Fig. 8-6. The difference in the values of I_{min} for $W_h = 0.076$ W may be explained by inaccuracy in establishing the value of I_{min}.

8.3. Transition of a Superconducting Solenoid into the Normal State

Investigations into uncontrollable processes involving the propagation of the normal zone were started at almost the same time as the creation of the first superconducting solenoids. This is because an accidental transition of the solenoid into the normal state usually puts it completely out of action. Repeated attempts have been made by one way and another to protect the solenoid from destruction. However, owing to the lack of information as to the processes taking place in a compact superconducting coil on passing into the normal state, all such methods have been of a unilateral nature, having been mainly directed at ensuring the rapid removal of stored energy from the solenoid. Systematic investigations into the transitional processes only began after 1963 [37, 38, 87].

In studying the mechanism underlying the transition of a close-packed coil into the normal state a great contribution was made by Stekly and his colleagues [88]. The creation of completely stabilized conductors cooled by liquid helium for a certain period reduced interest in close-packed coils. However, the appearance of internally stabilized conductors (and superconducting coils based on these) with high current densities restored attention to this problem.

As indicated in the preceding section, when a close-packed solenoid is operating in the current range corresponding to unstable states, the formation of a normal-zone nucleus in the coil may lead to the unintentional transition of the solenoid into the normal state.

In order to study the character of normal-zone propagation, as a first approximation we may use Stekly's proposed model for solenoids wound from a single superconducting conductor [88]. According to this model, the time dependence of the resistance of that part of the coil which has passed into the normal state is described by the following relationships, which respectively govern the one-, two-, and three-dimensional propagation of the normal zone:

$$
\left.
\begin{aligned}
R(t) &= \rho \int_0^t \mathfrak{v}_1 \, dt; \\[2mm]
R(t) &= (\beta\rho/d_{\text{con}}) \int_0^t \mathfrak{v}_1 \, dt \int_0^t \mathfrak{v}_2 \, dt; \\[2mm]
R(t) &= (2/3)(\beta\rho/d_{\text{con}}^2) \int_0^t \mathfrak{v}_1 \, dt \int_0^t \mathfrak{v}_2 \, dt \int_0^t \mathfrak{v}_3 \, dt,
\end{aligned}
\right\}
\tag{8-4}
$$

where β is the occupation factor of the coil relative to the superconducting material, ρ is the electrical resistance of unit length of the conductor in the normal state, d_{con} is the diameter of the conductor and $\mathfrak{v}_1$, $\mathfrak{v}_2$, $\mathfrak{v}_3$, are the rates of propagation of the normal zone along the conductor, from turn to turn in one layer (axial), and from layer to layer (radial).

When the solenoid is operating in the "frozen current" mode the transient process in the coil is described by the equation

$$
L \frac{d\iota}{d\theta} + \iota R(\theta) = 0,
\tag{8-5}
$$

where $\iota = I/I_0$ is the dimensionless current in the coil, I_0 is the initially "frozen current," $\theta = t/t^*$ is the dimensionless time, t^* is a certain function having the dimensions of time, and L is the inductance of the solenoid.

The solution of Eq. (8-5) takes the form

$$
\iota = \exp \left\{ - \int_0^\theta [t^* R(\theta)/L] \, d\theta \right\}.
\tag{8-6}
$$

This relation characterizes the time dependence of the attenuating current in the short-circuited coil when a normal-zone nucleus is created in the latter.

Clearly the voltage at the ends of the section which has passed into the normal state may easily be expressed in dimensionless form

$$\frac{t^*U}{\iota_0 L} = [tR\,(t)/L]\exp\left\{-\int_0^t [tR\,(t)/L]\,dt\right\}. \qquad (8\text{-}7)$$

The expression for the time constant depends on the character of normal-zone propagation:

for one-dimensional propagation

$$t^* = (L/\rho\mathfrak{v}_0)^{1/2}\,;$$

for two-dimensional propagation

$$t^* = \left(\frac{Ld_{\mathrm{con}}\mathfrak{v}_1}{\beta\rho\mathfrak{v}_0^2\mathfrak{v}_2}\right)^{1/3}\,;$$

for three-dimensional propagation

$$t^* = [3/2\,(Ld_{\mathrm{con}}^2/\beta\rho\mathfrak{v}_0^3\,)\,(\mathfrak{v}_1^2/\mathfrak{v}_2\mathfrak{v}_3)]^{1/4}, \qquad \left.\begin{matrix}\\[6em]\end{matrix}\right\} \qquad (8\text{-}8)$$

where $\mathfrak{v}_0$ is the rate of normal-zone propagation for an initial current I_0.

In analyzing Eqs. (8-6) and (8-7) we must remember that the rate of normal-zone propagation depends on the current and the magnetic field. As a first approximation the following relationship was proposed in [8-5]:

$$\frac{\mathfrak{v}_1}{\mathfrak{v}_0} = \frac{1 + \delta\,(B/B_0)}{1 + \delta}, \qquad (8\text{-}9)$$

where B_0 is the magnetic induction for $\iota = 1$ while δ is a constant close to unity which may be determined experimentally. Equation (8-9) may be converted to the form

$$\frac{\mathfrak{v}_1}{\mathfrak{v}_0} = \iota\,\frac{1 + \delta\iota}{1 + \delta}. \qquad (8\text{-}10)$$

Allowing for (8-10), Eq. (8-6) may be expressed more conveniently,

$$\frac{d\iota}{d\theta} = -\iota\left[\int_0^\theta \iota\,\frac{1 + \delta\iota}{1 + \delta}\,d\theta\right]^n, \qquad (8\text{-}11)$$

which makes it easy to estimate the number of directions of normal-zone propagation; here n = 1 for the one-dimensional, n = 2
for the two-dimensional, and n = 3 for the three-dimensional process.

We used the single, short-circuited solenoid method of Chapter
6 in order to make an experimental study of the principal laws
underlying normal-zone propagation in a close-packed coil. We
studied a solenoid with a close winding of seven-strand indium-
filled cable (six superconducting and one copper strand). The internal diameter of the solenoid was 30 mm, the external diameter
67.7 mm, the height 77 mm, and the inductance 0.029 H. The current
I_p^{cp} equaled 49 A.

The results of experiments carried out for various values of
the initial current in the solenoid I_0 are presented in Figs. 8-8 to
8-10. Figure 8-8 shows the current dependence of the voltage
drop U_{nz} in the part of the compact winding which has passed into
the normal state; Fig. 8-9 shows the current dependence of the
resistance R_{nz} of this section (the different curves of the dynamic
processes relate to different values of the initial current I_0). Figure 8-10 shows the current dependence of the instantaneous power
of heat evolution W.

We see from these figures that the character of the lines describing the dynamic process in the close-packed coil differs from

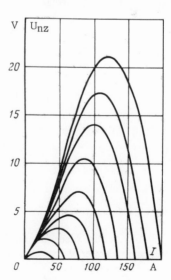

Fig. 8-8. Current dependence of the
voltage drop in the section of close-
packed coil which has passed into the
normal state.

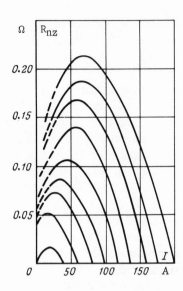

Fig. 8-9. Current dependence of the resistance of the section of close-packed coil which has passed into the normal state.

that of the same lines in the loose coil. Firstly, for $I = I_p^{cp}$ the lines of the dynamic process exhibit no singularities. This is because, in contrast to loose coils in which the dimensions of the normal zone diminish when the current falls below I_p (see, for example, Fig. 6-2), in the case of close-packed coils the points on the curves of the dynamic processes corresponding to the maxi-

Fig. 8-10. Current dependence of the instantaneous power of heat evolution in the close-packed coil.

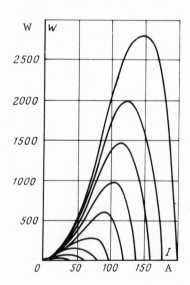

mum resistance of the normal zone should lie on the line $f = 0$ (see Fig. 8-1). Secondly, for any values of the initially frozen current, the residual currents in the coil are practically equal to zero after the attenuation of the dynamic process since in the case of close-packed coils the concept of a minimum current for the existence of the normal zone, at which the rate of "collapse" of this zone becomes extremely high, no longer exists.

When studying the basic laws of normal-zone propagation in a close-packed coil an important part is played by the number of directions in which this propagation occurs. We remember that in a loose coil the propagation of the normal zone is one-dimensional (it only moves along the combined conductor). In a close-packed coil (as indicated in Section 8.1) this process clearly has a three-dimensional character.

The question as to the number of directions of propagation actually characterizing a close-packed coil may be solved by analyzing experimental data regarding the time dependence of the coil current when a dynamic process is taking place in a short-circuited solenoid, using Eq. (8-11).

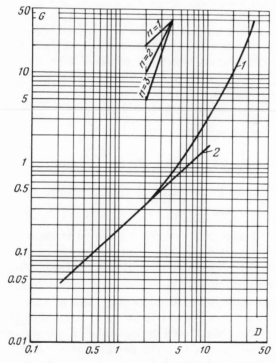

Fig. 8-11. The $G(D)$ relationship for a solenoid with a close-packed winding (curve 1) and a loose winding (curve 2).

If we possess sufficient experimental data regarding this relationship, we may calculate the quantities

$$D = \frac{1}{\iota} \frac{d\iota}{dt}$$

and

$$\left.\begin{array}{l} \\ G = \int_0^t \iota \left(\frac{1 + \delta\iota}{1 + \delta} \right) dt. \end{array}\right\} \qquad (8\text{-}12)$$

for various values of the defining parameters.

If we plot the values of D corresponding to each value of G in logarithmic coordinates, we may determine the index

$$n = \frac{d \ln G}{d \ln D}. \qquad (8\text{-}13)$$

Figure 8-11 gives the G(D) relationship for a solenoid with a close winding of seven-strand, indium-filled cable (curve 1) and also for the earlier-considered (Section 6.2) solenoid with a loose winding of the same cable (curve 2). The internal diameter of the solenoid with the close winding was 30 mm, the external diameter 71 mm, the height 77 mm, the number of turns 1440, and the inductance 0.037 H. The main parameters of the solenoid with the loose winding were given earlier.

Figure 8-11 also shows the theoretically calculated slopes of the curves for particular types of propagation (n = 1, 2, 3). We see from the figure that for a loose coil (as expected) n = 1, i.e., the propagation of the normal zone has a one-dimensional character. For a compact coil, at the onset of the dynamic process the propagation of the normal zone has a one-dimensional character, gradually changing into two- and finally three-dimensional.

8.4. Comparison of the Parameters of Transient Processes in Close and Thinly Packed Coils

It is of great practical interest to compare the parameters of normal-zone propagation in loose and densely packed superconducting coils made from partly-stabilized conductors. Since in such conductors the proportion of the cross section occupied by the normal stabilizing metal is relatively small, partly stabilized coils (both compact and loose) have a greater constructive current density than completely stabilized coils. When a normal-zone nucleus is formed, the coil passes into the normal state. In this connection it is important to establish which of the coils under comparison has the greater reliability in relation to possible burning out as

TABLE 8-2

Parameters of the solenoids	Compact coil	Loose coil
Internal diameter of winding, mm	30	30
External diameter, mm	67.7	82
Height of winding, mm	77	77
Number of turns	1200	1065
Magnetic constant of solenoid, T /A	0.0168	0.0140
Inductance, H	0.029	0.022

a result of Joule heat evolution in the normal zone. In addition to this it is essential to discover which of the coils has the greater reliability as regards the possible loss of electrical strength (breakdown of the insulation) due to the appearance of a large potential difference in the normal zone.

With a view to making a comparative analysis of this kind we set up some special experiments [89]. These experiments were carried out with two short-circuited solenoids made from a seven-strand indium-filled cable (six superconducting and one copper strand). The winding of one of the solenoids was made compact and that of the other loose. The solenoids were constructed in such a way as to make their parameters as nearly as possible equal; this was done so as to ease subsequent comparison of the results of the experiments. The principal characteristics of the solenoids are given in Table 8-2.

The experiments were carried out by the single-solenoid method for a helium tank temperature of 4.2 K. The initially "frozen" current equaled 150 A. The low degree of stabilization of the cable used as coil material was chosen in order to ensure zero residual currents. As a result of this, the energy liberated in the coil was roughly the same in both cases.

The results of the experiments are presented in Figs. 8-12 to 8-14. Figure 8-12 shows the time dependence of the instantaneous power of heat evolution W_{nz} in the section of winding which has passed into the normal state; Fig. 8-13 gives the time dependence of the resistance of the normal zone R_{nz}. We see from these two figures that the time constant of normal-zone propagation is slightly greater for the solenoid with the loose winding. The resistance of the normal zone, and also (in view of the fact that the time constants of the solenoids are not too different) the instantaneous pow-

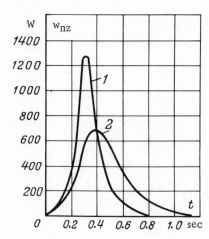

Fig. 8-12. Time dependence of the instantaneous power of heat evolution in the section of the coil which has passed into the normal state. 1) For the compact winding; 2) for the loose winding.

er of heat evolution, are considerably higher in the solenoid with the compact winding than in that with the loose winding. For the solenoid with the compact winding the maximum power of heat evolution exceeds 1.2 kW, which is a huge value for such a small solenoid.

It should be emphasized that the values of W_{nz} and R_{nz} by themselves are not so important when making a comparative analysis of the characteristics of the two kinds of superconducting coil under consideration. From the point of view of an analysis of the possible overheating of the coil as a result of heat evolution, it is

Fig. 8-13. Time dependence of the resistance of the normal zone. 1) For the compact winding; 2) for the loose winding.

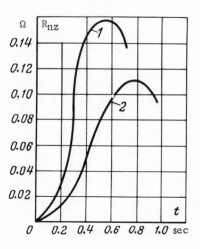

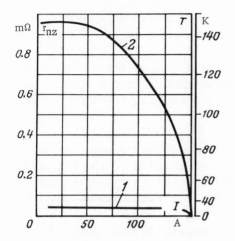

Fig. 8-14. Current dependence of the resistance of the auxiliary measuring section. 1) For a compact winding; 2) for a loose winding.

far more important to establish how these quantities are distributed over the coil – whether they are localized in a specific region or "spread" over the winding. The basic information in this respect was obtained from Fig. 8-14, which gives the current dependence of r_{nz}, the resistance of the auxiliary measuring section of the combined conductor (1 cm long). This relationship was obtained by calculation from the time dependence of the voltage drop U in the auxiliary measuring section recorded during the experiments.

In the center of the auxiliary measuring section we placed a diagnostic heater, to introduce a normal-zone nucleus into the coil. Thus this section, playing the part of a resistance thermometer, lay in the center of that part of the winding which was occupied by the normal zone, and clearly had the highest temperature.

We see from the curves of Fig. 8-14 that the resistance r_{nz}, and correspondingly the temperature in the center of the part occupied by the normal zone, is many times smaller in the compact than in the loose winding. In other words, during the propagation of the normal zone in the cooled (loose) winding a higher temperature is reached than in the uncooled (compact) winding. This unexpected result may be explained in the following way. The fact that the resistance of the normal zone R_{nz} in the compact winding is much greater than in the loose winding (Fig. 8-13), taken in conjunction with the data of Fig. 8-14, indicates that the total length of the section of combined conductor occupied by the normal zone is much greater in the compact than in the loose winding. Simple

calculations show that the length of that part of the cable which has
passed into the normal state is about 1 m for the loose and 100 m
for the compact coil.

It should be noted that the difference in the lengths of the non-
superconducting part arises from the basic factor distinguishing
normal-zone propagation in the loose and compact windings — in
the former it is a one- and in the latter a three-dimensional process.

Since the stored energy is approximately the same for the two
solenoids (the inductances being similar and the initial current I_0
identical), and since the time constants of the process in these
solenoids differ very little, it follows from the foregoing arguments
that the density of Joule heat evolution per unit length of the conductors is much lower in the compact than in the loose coil. The
same amount of energy is released over a much greater length in
the compact winding than in the loose.

The extent to which the temperature rises in the normal zone
is naturally determined not solely by the intensity of heat evolution
but also by the conditions of heat outflow into the surrounding medium. For that part of the conductor in the winding of the cooled
solenoid which at the time of the transient process is at the highest
temperature and lies in the center of the normal zone, heat flow
through the metal along the conductor may clearly be neglected.
In order to explain the relatively substantial heating of the normal
zone in the cooled winding we may therefore make use of the concepts developed in Chapter 5 for a combined conductor having a
constant temperature at every point.

We showed earlier that a characteristic current I* existed for
such a combined conductor, its value depending on the individual
properties of the conductor and the cooling conditions; above this
current, thermal equilibrium could not be achieved under any circumstances. Hence after the sample has passed into the normal
state its temperature rises unrestrictedly until the conductor burns
out.

The transition of the coil into the normal state for a current
$I_0 > I^*$ is accompanied by a rise in temperature, which under specified cooling conditions depends to a large extent on the stored
energy of the solenoid's magnetic field. For a solenoid of high
inductance ($I \approx$ const) overheating of the conductor is inevitable.
In small solenoids the current in the winding falls quite rapidly.

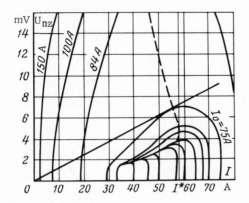

Fig. 8-15. Volt–ampere characteristics of part
of the conductor corresponding to the transition
of the solenoid with a loose winding into the normal state.

A certain maximum temperature is then reached in the winding.
The maximum temperature depends on the inductance of the solenoid and on the difference $\Delta I = I_0 - I^*$.

Figure 8-15 shows the volt–ampere characteristics of part of
the conductor corresponding to the transition of the solenoid under
consideration (with the loose winding) into the normal state for
various values of I_0. Using the method of Section 6.2, we may
analyze the foregoing transition curves so as to restore the volt–
ampere characteristic corresponding to the states of thermal equilibrium of this particular conductor (broken line). It follows from
the figure that for the conductor under consideration $I^* \approx 57$ A,
and hence the transition of the loose coil for an initial "frozen
current" of $I_0 = 150$ A should be accompanied by a considerable
rise in temperature.

We see from Fig. 8-14 that, in the small experimental solenoid with the loose winding considered, a temperature of about
150 K is reached during the propagation of the normal zone. Of
course, in much larger superconducting magnetic systems with
loose coils, the rise in temperature during the propagation of the
normal zone may be so great that (in the absence of protection) it
will lead to the burning out of the winding.

It is thus all too clear from the foregoing that, paradoxical
as it may seem at first glance, the cooled (loose) windings may be

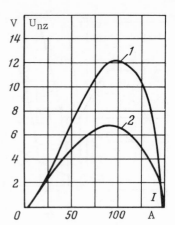

Fig. 8-16. Volt—ampere characteristics of part of the winding which
has passed into the normal state. 1)
For a compact winding; 2) for a loose
winding.

less reliable with respect to burning out than the uncooled (compact) windings.

Figure 8-16 shows the potential difference in the part of the coil which has passed into the normal state, U_{nz}, as a function of the current I in the coil. We see from the figure that the maximum value of U_{nz} for the compact winding is almost twice as great as that of the loose winding. This is because, as already noted, the value of R_{nz} for the compact winding is much higher than that of the loose winding. Furthermore the inductance of the experimental solenoids under consideration is practically the same (see Table 8-2), while the time constants of the normal-zone propagation process are not very different.

It should be emphasized that the small value of U_{nz} determined for the solenoids under consideration does not mean that during the passage of the normal zone along the superconducting coil the value of U_{nz} will not reach dangerous values. In the present case we have only considered miniature experimental solenoids intended for studying qualitative laws. Practice shows that in large systems the potential difference U_{nz} increases very sharply; even in relatively small experimental solenoids with an inductance of only a few henries the voltage reaches 500-1000 V.

It is accordingly quite obvious that compact windings are less reliable than loose ones, particularly relative to the possibility of electrical damage (breakdown of the insulation) during the propagation of the normal zone.

Thus superconducting magnetic systems with compact winding should be constructed with due allowance for the necessity of protecting them, first of all, from excess voltage during the transition into the normal state. The question of protecting windings of this type from excessive heating due to Joule heat evolution is of secondary importance.

Part III

COMBINED CONDUCTORS WITH INTERNAL STABILIZATION

Chapter 9

Stability of Superconductors of the Second Kind with Respect to Flux Jumps

9.1. General Principles

In Part II we considered the basic principles of the methods employed for the thermal (cryostatic) stabilization of superconductors. This method ensured the safe and reliable operation of a superconducting magnetic system in the presence of certain acceptable perturbations. Only serious mechanical and electrical damage or the stoppage of the coolant can lead to failure or disorder in such a system. However, ensuring a high degree of reliability by using the method in question involves seriously reducing the mean efficiency of the current density in the windings, and for a number of systems this may be quite unacceptable.

Methods of producing combined superconducting materials with so-called internal stabilization have been developed over the last few years.* This term, which is not entirely satisfactory, should only emphasize the distinction from cryostatic stabilization, in which the result of the perturbation is suppressed by means of additional material, external to the superconductor. Internal stabilization is directed toward eliminating the very possibility of at least one particular form of perturbation characteristic of nonideal superconductors of the second kind (flux jumps), or to suppress their development by virtue of internal processes.

In this part we shall go into more detail than we did in Section 2.3 and consider both the reasons for the development of flux jumps

---

*Other terms used (not entirely equivalent to the foregoing) are "intrinsically stable," i.e., stable by itself, or filamentary, i.e., using filamentary conductors.

and also the special measures capable of being taken to exclude or reduce these. We shall also consider certain additional problems associated with the use of stabilized conductors in pulsed magnetic systems.

It should be noted that ensuring internal stabilization, i.e., eliminating flux jumps, does not guarantee the system against the action of other undesirable perturbations. Thus in such systems additional measures must be taken to protect against actions which might even partly vitiate the advantages of internal stabilization. The most typical example of this is the effect arising from the movement of poorly fixed coil windings. The heat caused by this movement may be sufficient to initiate a transition of the magnetic system into the normal state. Hence ensuring mechanical stability is a vital problem in such systems. At the present time this problem is being attacked by purely empirical methods without regard to the scientific principles underlying them, the development of which has only just started; we shall therefore deal with this question only briefly. Although particular problems in this respect may be successfully solved, to ensure stability with regard to all conceivable perturbations (vibrations and so forth) is hardly feasible. Hence any magnetic system based on conductors with internal stabilization should in all cases be designed (assuming that the system is not intended to be used once only) in such a way as to allow it to pass harmlessly into the normal state (see Section 3.1). In a number of cases additional external means of protection may be needed to ensure this.

9.2. Magnetization of Nonideal Superconductors of the Second Kind

As indicated in Chapter 2, in superconductors of the second kind the magnetic field is able to penetrate into the interior of the material. The character of the magnetization in ideal (i.e., free from all defects) and nonideal superconductors (the latter sometimes being called "hard" for the sake of brevity) is therefore quite different (see, for example, Fig. 2-7). The reason for the difference lies in the fact that hard superconductors are able to pass considerable currents in a direction perpendicular to the direction of the magnetic field; it is this property which enables them to be used in producing strong magnetic fields.

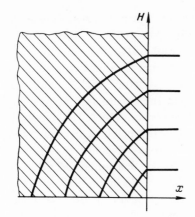

Fig. 9-1. Magnetic-field distribution in the interior of a superconductor.

Let us consider an idealized, isothermal, i.e., quasi-stationary, process of magnetization in a nonideal superconductor of the second kind. Let the superconductor occupy a space x < 0 (Fig. 9-1). On increasing the external magnetic field H_e above the value $H_{c\,1}$ the field penetrates into the interior of the material; the change in electric field so created excites screening currents in the material, directed parallel to the surface of the superconductor and perpendicular to the external field.

Let us assume that after the field reaches a certain value its rise stops. The screening currents induced up to this instant will become attenuated; in each cross section the current will tend toward the critical value. This state of the superconductor is called the critical state. The steady magnetic-field distribution is given by the Maxwell equation, which takes the following simple form in the case considered

$$\frac{\partial H}{\partial x} = J_c\,(H). \qquad (9\text{-}1)$$

The $J_c(H)$ relationship is determined by the properties of the particular material. Since the critical current density usually diminishes with increasing field, the resultant field distribution has a negative curvature. The field distribution in the interior of the superconductor is shown in Fig. 9-1 for several successively increasing values of the external field.

Let us consider a thin infinite layer of superconductor 2b thick (Fig. 9-2). When the field increases on both sides of the layer, it

will gradually penetrate into the sample; for a certain value H_p
the magnetic field will penetrate into the middle of the sample, and
then the field in the center of the sample will gradually start increasing. We note that when the field has attained values greatly
exceeding H_p the curvature of its distribution may be neglected in
any practical calculations, or (amounting to the same principle)
there will no longer be any need to consider the J_c (H) dependence,
i.e., we may consider that the field distribution is represented by
two sections of straight lines with slope $J_e(H_e)$.

The break in the field distribution in the center of the sample
corresponds to the fact that at this point the currents abruptly
change direction by 180°. This change in current direction cannot,
of course, occur in a superconductor of the second kind at distances
less than the coherence length ξ. However, since this quantity is
extremely small in hard superconductors (Section 2.2), for any
macroscopic problems it is perfectly satisfactory to assume that
the current changes abruptly.

Let us estimate the additional magnetic energy associated
with the screening currents for the layer illustrated in Fig. 9-2.
Since we are only interested in that part of the total magnetic energy which may vary on varying the current distribution, we should
only calculate the magnetic-field energy arising from these currents, not counting the fields of the external sources. We may furthermore neglect the influence of the equilibrium magnetization of
the superconductor, since this becomes extremely small in the re-

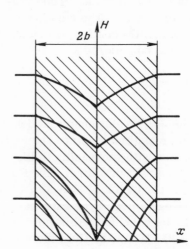

Fig. 9-2. Penetration of the magnetic field inside a thin superconducting layer.

gion of strong fields which is of greatest present interest. Thus the relationship for the energy density of the field due to the currents takes the following simple form:

$$\mu_0 \, \frac{H^2}{2} = \begin{cases} 1/2\mu_0 J_c^2 (x+b)^2 & \text{for } x < 0; \\ 1/2\mu_0 J_c^2 (x-b)^2 & \text{for } x > 0. \end{cases} \qquad (9\text{-}2)$$

Integrating this we obtain the total energy for the specified current distribution, referred to unit length and height of the layer:

$$\varepsilon = \frac{\mu_0}{3} J_c^2 b^3. \qquad (9\text{-}3)$$

The proportion of magnetic energy associated with the screening currents in the superconductor is, strictly speaking, negative, since the total magnetic energy incorporated within the volume of the superconductor diminishes in the presence of these currents. The purpose of the calculation is to estimate the energy which may be liberated within the superconductor in the form of heat when there is a change in the screening currents, especially when these die out altogether. In such cases the energy losses and the change in the resultant magnetic energy are made up by the work of the source creating the external magnetic field.

For current densities J_c characteristic of present-day hard conductors (J_c is usually greater than 10^9 A/m^2), the energy of the screening currents may reach rather large values, entirely compatible with the total internal energy of the superconductor. This high energy of the currents creates a source of instability relative to their distribution in the superconductor. The losses taking place during fluctuations of the screening currents may lead to an increase in the temperature of the conductor, which reduces the local critical current density, i.e., cause additional losses, and so on. The instability so developing appears in the form of a flux jump, i.e., a sharp redistribution of the induction in a certain part of the winding. As indicated in Chapter 8, the flux jump may cause the winding to pass completely into the normal state; i.e., it constitutes a cause of "degradation," an undesirable reduction in the limiting permissible current of the magnetic system.

We see from Eq. (9-3) that the mean energy density of the screening currents, i.e., the energy referred to a unit volume of

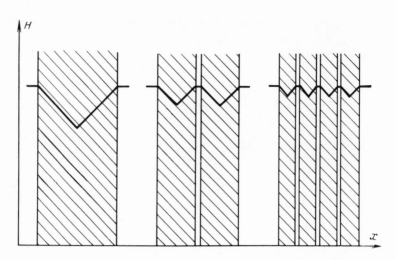

Fig. 9-3. Changes in the distribution of the magnetic field on making the superconductor successively thinner.

the superconductor, increases in proportion to the square of the thickness of the latter. This leads to one of the most important principles underlying methods of internal stabilization: In order to reduce the influence of the screening currents, i.e., in order to eliminate the instabilities associated with their action, it is essential to use a superconductor in the form of fairly thin layers (or thin wire). The reduction in the total energy of the screening currents as the conductor is made progressively thinner is clearly illustrated in Fig. 9-3, which represents the field distribution corresponding to the magnetization of the original layer and sets of thin isolated layers of the same material amounting to the same total thickness. We see that for a sufficiently fine division of the superconductor the energy of the screening currents may be reduced to quite small values.

These simple considerations were developed as early as 1964-1965 and were first proposed by Chester [90]. In the scientific literature the principles of internal stabilization only started being discussed later [45, 91, 92], when the possibility of manufacturing fine filamentary* or "whisker" conductors was noted.

---

*We should not confuse this term with the notation sometimes also used for spongy materials forming a matrix containing fine, interwoven, mutually interconnected channels, filled with a superconductor.

The production of wire (with a diameter of the order of $1-10\,\mu$) by ordinary technology, i.e., drawing, is difficult and expensive. The winding and use of coils made from such wire are also extremely difficult (the coils would have a very high inductance and low reliability). Hence the internally stabilized conductors now being produced are of the multicore type, i.e., they constitute a set of large numbers of very fine superconducting filaments enclosed in a common metallic matrix. As we shall subsequently show, the properties of such a system of connected parallel superconductors are not completely identical with the properties of individual, isolated filaments. The special characteristics of multicore systems will be considered in Chapters 10 and 11; at the moment we shall consider certain criteria by which we have to be guided in choosing the parameters of the conductors required to ensure stability. For this purpose we shall consider the question of the stability of the current distribution in a single superconductor, only allowing for its interaction with the surrounding sheath of normal metal where this is necessary.

9.3. Stability of the Screening Currents in a Superconductor of the Second Kind

Earlier (Section 9.2) we presented a qualitative analysis of the development of instability in the current distribution of a superconductor, appearing in the form of a flux jump. For an exact quantitative investigation into this process it is essential to establish the range of variation in the parameters of the material for which the development of such an instability is possible.

In its strictest presentation such a problem is extremely complicated. The behavior of the superconductor is described by a system of two coupled differential equations for the temperature and current; owing to the rapidly varying electrical conductivity in the resistive state of the superconductor, the equation for the current is very nonlinear. On the other hand the parameters characterizing the thermophysical and electrophysical properties of the majority of superconductors employed have not as yet been determined with sufficient accuracy. All this entirely justifies the customary simplified approach to the problem in question.

Let us consider the field distribution in a thin layer of thickness 2b (Fig. 9-4). We assume that all the quantities are indepen-

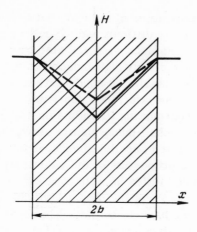

Fig. 9-4. Distribution of the magnetic field in a thin layer of superconductor.

dent of the coordinates y and z, i.e., the problem is essentially
one-dimensional in character. The heat-conduction equation for
the present case with the heat sources disconnected will take the
form

$$c \frac{\partial T}{\partial t} = \lambda_s \frac{\partial^2 T}{\partial x^2}, \qquad (9\text{-}4)$$

where c is the specific heat (heat capacity) per unit volume of the
conductor.

The diffusion equation of the electric field (magnetic field or
current) has an analogous form, well known from the theory of surface effects:

$$\frac{\partial E}{\partial t} = \frac{\rho}{\mu_0} \frac{\partial^2 E}{\partial x^2}, \qquad (9\text{-}5)$$

where E is the electric field strength.

We may simplify the problem by comparing the characteristic
times required for the establishment of thermal equilibrium τ_{te}
and for the penetration of the field into the sample τ_{em} :

$$\begin{cases} \tau_{te} \approx \dfrac{4}{\pi^2} \dfrac{b^2 c}{\lambda_s} ; \\[2mm] \tau_{em} \approx \dfrac{4}{\pi^2} \dfrac{b^2 \mu_0}{\rho} ; \\[2mm] \dfrac{\tau_{te}}{\tau_{em}} = \dfrac{c \rho}{\lambda_s \mu_0}. \end{cases} \qquad (9\text{-}6)$$

The slight indeterminacy in these relationships lies in the fact that ρ is here the resistivity ρ_f of the superconductor in the resistive state, which itself depends on the field (or current), varying from zero to a value close to the resistance in the normal state. If for purposes of estimation we take this latter (certainly overestimated) value, we find that the τ_{te}/τ_{em} ratio for all materials used should be 10^3-10^4 [93]. It is quite clear that for sufficiently rapid processes, in which substantial electric fields develop in the material, the ratio of the characteristic times will tend toward the larger value indicated, and only for extremely slow processes in which ρ_f decreases to zero will this ratio also diminish. Thus it is entirely permissible to assume that in the resistive state the field becomes established more quickly than temperature equilibrium is reached. It may also be shown that such an assumption yields a stability criterion which incorporates a certain reserve. If in actual fact the heat is able to flow away from those regions in which it is initially liberated, the operating conditions of the superconductor are only made easier. Naturally in this case we need not allow for the influence of the sheath surrounding the superconductor — we may regard the superconductor as existing under conditions of adiabatic insulation.

On presenting the problem of the stability of the critical state in this way, a fairly rigorous solution may be obtained. Such solutions were first obtained by Wipf [94] and Swartz and Bean [95], who considered the character of the equilibrium for small deviations of the external field from its original value.

We shall now give a slightly simplified derivation of the stability criterion [93], and also note the principles of a more rigorous solution based on studying the development of slight temperature perturbations. We shall assume that an infinitely small temperature jump ΔT_0 takes place throughout the whole thickness of the superconductor. As a result of this the critical current density decreases by an amount

$$\Delta J_c = \frac{\partial J_c}{\partial T} \Delta T_0 \tag{9-7}$$

and the field undergoes a certain redistribution, now being determined by a new value of the critical current (Fig. 9-4). Let us calculate the heat loss density associated with the redistribution

$$\Delta q\,(x) = \int JE\,dt \approx J_c\,\Delta\Phi = -\,J_c\mu_0\,\Delta J \left|\left(b - \frac{x}{2}\right)x\right| =$$
$$= -\,\mu_0 J_c\,\frac{\partial J_c}{\partial T}\left|x\left(b - \frac{x}{2}\right)\right|\Delta T_0. \tag{9-8}$$

We see from this relationship that heat evolution takes place in a nonuniform manner over the cross section, in such a way that the loss intensity decreases to zero in the center of the sample. This may easily be understood if we remember that the electric field and current density change sign in this region.

For a further approximation we average the losses over the whole cross section

$$\overline{\Delta q} = \frac{1}{b}\int_0^b \Delta q\,(x)\,dx \approx -\frac{\mu_0}{3}\,b^2 J_c\,\frac{\partial J_c}{\partial T}\,\Delta T_0. \tag{9-9}$$

We note that the average losses are equal to the change in the energy of the screening currents. Thus as a result of the field redistribution the temperature undergoes a further rise by an amount

$$\Delta T_1 = \frac{\overline{\Delta q}}{c} = -\frac{\mu_0}{3c}\,b^2 J_c\,\frac{\partial J_c}{\partial T}\,\Delta T_0. \tag{9-10}$$

This heating in turn leads to a new redistribution of the fields and hence further heating, and so on. The final temperature after several such steps equals the sum of the terms of a geometrical progression with a denominator determined by (9-10).

If this quantity is no greater than unity, the process should be convergent. Hence we obtain the following very important criterion of adiabatic stability:

$$b^2 \leqslant -\frac{3c}{\mu_0 J_c\,\frac{\partial J_c}{\partial T}}. \tag{9-11}$$

Sometimes this inequality is presented in the following more convenient form:

$$bJ_c \leqslant \left(\frac{3cT_0}{\mu_0}\right)^{\frac{1}{2}}, \tag{9-12}$$

where

$$T_0 = -J_c \left/ \frac{\partial J_c}{\partial T} \right. . \tag{9-13}$$

On the basis of Eq. (4-3), for the majority of materials

$$J_c(T) \approx J_c(T_t) \left(1 - \frac{T - T_t}{T_{c0} - T_t} \right), \tag{9-14}$$

and we therefore have

$$T_0 \approx T_{c0} - T_t . \tag{9-15}$$

Criterion (9-12) may be expressed in another form by introducing the concept of the screening field $H_p = bJ_c$:

$$H_p \leqslant \left(\frac{3cT_0}{\mu_0} \right)^{\frac{1}{2}} . \tag{9-16}$$

The foregoing derivation of the criterion of adiabatic stability contains some assumptions not entirely consistent. Thus it is not quite clear whether it is legitimate to represent the process in a stepped manner, nor is it certain whether convergence may validly be asserted. Furthermore, in contrast to the assumption that no heat transfer occurs in the sample, all the heat liberated is actually distributed uniformly over the cross section. However, these assumptions only lead to very slight deviations from the stricter result, amounting to a slightly different numerical coefficient in the final equation (9-11). A strict derivation of the exact criterion may be carried out, for example, as follows.

According to the foregoing assumptions the current distribution is able to "follow" the temperature distribution, i.e., we always have the relationships

$$J = J_c + \frac{\partial J_c}{\partial T} \, \Delta T. \tag{9-17}$$

The power of the heat liberated may therefore be written

$$\dot{q}(x) = JE \approx -J_c \dot{\Phi} = J_c \int_0^x \frac{\partial B}{\partial t} \, d\xi = J_c \mu_0 \int_0^x d\xi \, \frac{\partial}{\partial t} \int_\xi^b J(\eta) \, d\eta =$$

$$= J_c \mu_0 \int_0^x d\xi \int_0^\xi d\eta \, \frac{\partial J_c}{\partial T} \frac{\partial T}{\partial t} . \tag{9-18}$$

From this we easily obtain the following heat-conduction equation for a small temperature increment $\beta(x, t)$

$$c \frac{\partial \beta}{\partial t} = \lambda \frac{\partial^2 \beta}{\partial x^2} + J_c \mu_0 \int d\xi \int d\eta \frac{\partial I_c}{\partial T} \frac{\partial \beta}{\partial t} . \qquad (9-19)$$

Executing a twofold differentiation with respect to x, we have

$$c \dot{\beta}'' = \lambda \beta^{IV} - \frac{\mu_0 J_c^2}{T_0} \dot{\beta}. \qquad (9-20)$$

The development of the perturbation $\beta(x, t)$ is then analyzed in the same way as in Section 5.5, although in the present case the situation is rather more complex. Let us express β in the form of a series with respect to certain harmonics, subsequently to be determined:

$$\beta = \sum_\Lambda T_\Lambda(t) X_\Lambda(x). \qquad (9-21)$$

Substituting one of the terms of this series into the original equation and transforming, we obtain

$$\frac{\dot{T}_\Lambda}{T_\Lambda} = \frac{\lambda \dfrac{X_\Lambda^{IV}}{X_\Lambda}}{c \dfrac{X''_\Lambda}{X_\Lambda} + \dfrac{\mu_0 J_c^2}{T_0}} = \Lambda. \qquad (9-22)$$

Here as usual we have equated the two terms to a certain constant Λ, since the two sides of the equation are functions of different variables. The Λ spectrum is determined by the equation

$$\lambda X_\Lambda^{IV} - \Lambda \left[c X''_\Lambda + \frac{\mu_0 J_c^2}{T_0} X_\Lambda \right] = 0. \qquad (9-23)$$

The boundary conditions at the ends of the section under consideration are determined by the adiabatic conditions

$$X'_\Lambda(b) = X'_\Lambda(-b) = 0. \qquad (9-24)$$

To this we must add certain extra conditions. At the point x = 0 the heat evolution vanishes, which is automatically taken into account in the integrodifferential equation (9-19); however, it is in no way reflected in Eq. (9-20), since the constant terms in this vanished after differentiation. At the point x = 0 we have the equation

$$c \dot{\beta} = \lambda \beta'', \qquad (9-25)$$

whence

$$\frac{\dot{T}_\Lambda}{T_\Lambda} = \frac{\lambda}{c} \frac{X''_\Lambda}{X_\Lambda} = \Lambda. \qquad (9-26)$$

Thus the condition at the point x = 0 takes the form

$$X''_\Lambda (0) = \Lambda \frac{c}{\lambda} X_\Lambda (0). \tag{9-27}$$

In an analogous manner we may obtain a condition for the point x = b:

$$X'''_\Lambda (b) = \Lambda \frac{c}{\lambda} X'_\Lambda (b). \tag{9-28}$$

These boundary conditions are sufficient for determining the spectrum of eigen-values Λ. A positive eigenvalue corresponding to the instability of the ground harmonic first appears on satisfying the condition

$$bJ_c = \left(\frac{\pi^2 c T_0}{4\mu_0} \right)^{\frac{1}{2}}. \tag{9-29}$$

Thus the exact solution leads to an expression differing from the results of the approximate calculation by no more than $2\sqrt{3}/\pi \approx 1.103$ times. The solutions proposed in [94, 95] also lead to Eq. (9-29). The exact calculation of the numerical coefficient in the expression for the stability criterion is by itself not especially important, since a reliable experimental verification of the criterion is quite difficult. However, by using the method in question we may in a number of cases estimate the influence of the substance in which the superconductor is immersed (and also the role of the finite magnetic-field diffusion velocity). When the surrounding medium is a dielectric (helium), the analysis is particularly simple, since here it is sufficient to change the boundary conditions in order to allow for heat transfer to the external medium.

By changing condition (9-28), in fact, we may also allow for the electromagnetic action of the currents in the sheath of normal metal surrounding the superconductor. A change in the conditions of heat transfer at the boundaries of the superconductor per se has no effect on the stability criterion (9-29), since the earliest loss of stability occurs in the case of rapid jumps in which the heat transfer is unable to exert any major influence. However, for a certain (fairly small) thickness of the normal sheath (because of its damping action) instability first appears for slow jumps, and the influence of the heat-transfer conditions then becomes considerable. Without delving further into this problem, we shall here simply give the expression for this critical thickness of the sheath

$$d_{cr} = \frac{8cb}{315\lambda_s \sigma_n \mu_0}, \tag{9-30}$$

where σ_n is the conductivity of the normal metal.

We should note one difficulty of a fundamental character in formulating the stability problem. The solutions obtained by our proposed method only give a reasonably accurate description of the motion of the perturbation when the temperature is rising at all points. If at any point the temperature tends to decrease, then in this region Eq. (9-17) ceases to be valid, since on cooling the current does not increase but "freezes," i.e., is fixed at the critical value. For an unstable perturbation in the form of a ground harmonic (possessing no zeroes), no difficulties arise; however, if the original perturbation contains higher harmonics, the motion becomes more complicated and its final result is not always clear. In every case the resultant solution signifies the onset of instability with respect to a perturbation of some specific form.

Let us now make some numerical estimates of the critical size of the superconductor 2b below which no flux jumps should occur. The specific heat of the majority of alloys at $T \approx 4$ K is usually a few $J/(dm^3\text{-}K)$. For $c = 3$ $J/(dm^3\text{-}K)$, $J_c = 5 \times 10^5$ A/cm^2, and $T_0 \approx 4$ K, Eq. (9-12) gives $2b \approx 30$ μ. The screening field is 1.7×10^5 A/m. We note that this value in no way depends on the current density taken or the critical size.

The experimental data regarding flux jumps on the whole agree qualitatively with our estimates. The manner in which the critical screening field $H_{s.cr}$ depends on the parameters of the material (specific heat) is also consistent with (9-16). It is difficult to establish exact quantitative agreement with this equation from existing experimental data, especially because the specific heats of the materials studied are usually not known with sufficient reliability, while the conditions of adiabatic insulation are not always reliably ensured. In any case we may consider that Eq. (9-12) provides a basis for quantitatively estimating the stability limits to within a factor of 2 to 3.

The stability criterion with respect to flux jumps expressed by Eq. (9-12) enables us to specify various ways in which stability of the superconductor may be ensured. The chief of these is reducing the size of the superconducting wires as already indicated. We may also attempt increasing the specific heat of the material artificially (for example, by working at higher temperatures). Also of practical importance is the saturation of porous Nb_3Sn samples with liquid helium, which has a relatively high specific heat. Certain possibilities opening in this direction will be considered in more detail in Section 10.2.

Reducing the critical current density of the sample, as indicated by Eq. (9-12), also increases its stability. This agrees closely with the well-known fact that close to the upper critical field B_{c2} no flux jumps are encountered when the current diminishes. Although by itself a reduction in the critical current density of the material may seem irrational, the optimum current density (largely determined by the conditions of mechanical and heat treatment) should of course be chosen with due allowance for the possibility of its stable realization in the coil. "Hard" materials such as Nb $-$ Zr alloys, commonly subjected to considerable deformation and having extremely high levels of current density (over 5×10^5 A/cm^2), are usually characterized by fairly low currents in the windings (approximately 2×10^4 A/cm^2). At the same time softer materials with a current density of under 10^5 A/cm^2 enable us to realize similar densities (referred to the cross section of the superconductor) fairly reliably.

In order to increase the stability of the magnetic system it is often recommended that a coil sectionalized with respect to radius and having a particular order of connecting the sections should be employed. The outer sections, which first come into operation, create a so-called "backing field" which ensures stable introduction of the inner sections. It is not difficult to see that a reduction in the current density of the central sections, situated in the high field of the outer sections, exerts an appreciable influence in this case. If the inner sections are brought into action in a low backing field or no field at all, flux jumps may occur in the zone of moderate fields, in which the current density is still high, and these jumps will lead to the premature transition of the system into the normal state.

Let us consider the influence of the derivative $\partial J_c / \partial T$ in Eq. (9-11) on stability. Naturally greater stability follows a reduction in this quantity. If we were able to develop a material for which $\partial J_c / \partial T > 0$, flux jumps would not occur for any thickness of the superconductor. Certain interesting possibilities in this field were considered by Livingstone [96]. It was found that after certain heat treatment Pb$-$In$-$Sn alloys underwent phase transformations with the precipitation of a finely dispersed second phase, also superconducting, but with a lower critical temperature. Thus for a sufficiently low temperature the superconductor is, as it were, homogeneous. On raising the temperature the second phase passes into the normal state and its particles start playing the role of detaining centers, as a result of which over a certain range of temperatures the critical current rises. As expected, the flux jumps encountered in other temperature ranges were completely absent in this case. The development of materials with such properties for reasonably strong fields is an extremely attractive possibility.

Let us now give some more detailed attention to the assumptions which we made when deriving the stability criterion. The one-dimensional model chosen is only reasonably representative of a strip-type conductor situated in a part of the coil in which the field is parallel to its plane. In the presence of appreciable field components perpendicular to the plane of the strip, the role of critical size should be played by a quantity of the order of the strip width, and in this case the problem becomes much more complicated. On the other hand, it is quite clear that the resultant criterion should be entirely suitable for estimating the critical diameter of a wire (the continuous layer which we are considering may be represented as consisting of individual turns). However, the interaction of the individual sections of the conductor and the neighboring turns cannot be taken into account in the one-dimensional presentation of the problem.

We note in this connection that there are many experimental results indicating that, during the attenuation of the screening currents (due, for example, to initial heating), the residual magnetization of the conductor is not uniform with respect to length but has a sharply periodic structure [9 7]. It is reasonable to assume that the attenuation of the currents is accompanied by the development of an instability which cannot be revealed in the one-dimensional model. Cases of a more complicated configuration of the conductor have as yet never been fully studied.

It should be remembered that in none of the calculations so far carried out has the so-called magnetocaloric effect been considered. This effect occurs for all materials in which the magnetization depends on the temperature; it amounts to the fact that when the magnetization alters, heat is emitted or absorbed. If the sample is adiabatically insulated, a change in magnetization causes a rise or decrease in temperature. The sign and magnitude of the effect when the external field varies under isothermal conditions may be determined from the Maxwell thermodynamic equation

$$\Delta q = - T \left(\frac{\partial S}{\partial H}\right)_T \Delta H = - T \left(\frac{\partial M}{\partial T}\right)_H \Delta H. \qquad (9\text{-}31)$$

Since superconductors in the superconducting state obey the equation $(\partial M/\partial T)_H > 0$ (M < 0), a rise in magnetic field causes the absorption of heat.

Thus this mechanism should in principle have a favorable stabilizing influence during magnetization. However, numerical estimates show that the effect is extremely slight in hard superconductors. This is in particular because the equilibrium magnetic moment which determines the thermodynamic functions of the superconductor is small (in no case does it exceed B_{c1}).

On the other hand, the specific losses associated with magnetization are in accordance with (9-9) proportional to the square of the thickness of the superconductor, while the magnetocaloric effect is proportional to the change in the screening field, i.e., to the first power of the thickness. Hence on reducing the thickness the role of this effect will increase. We may approximately consider that the influence of this effect will become appreciable when the screening field $H_p = bJ_c$ becomes comparable with B_{c1}/μ_0 for the particular material. In practice this condition may be satisfied for conductors a few microns thick, such as are used, for example, in pulse-type superconducting magnetic systems.

It should be noted that when the field varies cyclically the mean heat evolution in the superconductor should not depend very greatly on the magnetocaloric effect, which is thermodynamically reversible; there may still be an influence on the losses as a result of nonlinear effects.

In considering the development of initial perturbation it was assumed that the original state of the sample corresponded to an isothermal critical state in which the current density at each point was equal to the critical value for the particular temperature. However, as already noted, during the establishment of equilibrium, owing to the poor thermal conductivity of the material, the temperature of the superconductor is always slightly greater than the external temperature, so that the current density tends to set at a lower level. Thus a purely isothermal critical state can only be established as a limiting state for an infinitely slow rise in the external field.

The reserve of stability so created, however small it may be, affects the character of the breakdown of equilibrium. The point is that infinitely small perturbations (for example, thermal fluctuations) cannot break the established equilibrium, which thus becomes metastable rather than absolutely unstable. Hence a flux jump can only develop for fairly strong initial perturbations or for a not too slow change in the external conditions. Only such perturbations as embrace large parts of the superconductor (comparable with the critical dimension) will be able to lead to flux jumps, while the growth of small-scale fluctuations will be impeded.

Thus the infringement of criterion (9-12) still does not mean that flux jumps will immediately break the established metastable equilibrium state. "Starting mechanisms" initiating jumps may be generated by mechanical, thermal, and electromagnetic interaction of the coil turns.

9.4. Criterion of Adiabatic Stability

in the Presence of a Transport Current

In the preceding section we considered a model with screening currents only, the total transport current in the conductor being zero. Far more important in practice is the case in which a finite current flows in the conductor, because flux jumps may create a resistance to the transport current and initiate its attenuation.

In order to allow for the influence of a transport current on the stability condition of the conductor we must first of all establish the form of field distribution in the presence of a current. Let us return to the earlier simplified model with a plane superconducting layer. The field distribution within the layer depends on how the current and external field are increased.

If the original state of the sample is characterized by the complete penetration of the field (for example, if the sample is first heated to the normal state and then cooled in an external field) an

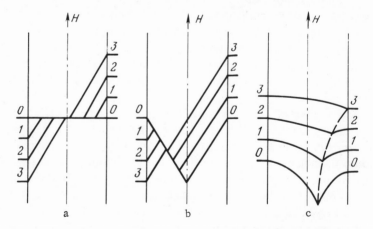

Fig. 9-5. Distribution of magnetic field inside the superconducting layer for various conditions of changing current and external field.

increase in current will lead to a gradual inward penetration of the field from both sides of the layer (Fig. 9-5a). The achievement of the critical transport current corresponds to the complete filling of the cross section, in which the field distribution is represented by a continuous straight line. The difference in field values on the two sides of the layer then becomes $2bJ_c = I_c$, i.e., equal to the critical current associated with unit height of the layer. If the current is increased after the external field has been fixed, the character of the distribution will be quite different (Fig. 9-5b). Another important case, in which the current and field are increased at the same time (as usually occurs in the coil of a magnetic system) is shown in Fig. 9-5c.

We note that in all these cases a line on which $E = 0$ and $J = 0$ exists in the interior of the superconductor up to the instant of reaching the critical current; on this line the field distribution experiences a break. In the last case which we considered this line is displaced in the material, as indicated by the broken curve of Fig. 9-5c. On reaching the critical current the line passes outside the layer.

Let us carry out a simplified calculation of the stability of the isothermal distribution shown in Fig. 9-5c. The basis of the calculation corresponds closely to the derivation of the criterion (9-12) reproduced in the preceding section. The only difference lies in the extra losses associated with the redistribution of the field. In order to determine the losses we may once again calculate the electric field E from the Maxwell equation

$$\frac{\partial E}{\partial x} = -\dot{B}. \tag{9-32}$$

For an infinitely small change in current, the displacement of the line on which $E = 0$ may be neglected and we may integrate Eq. (9-32) starting from this line. In this way we obtain the following relationship for the density of the losses:

$$\Delta q\,(x) = \begin{cases} -\mu_0 J_c \frac{\partial J_c}{\partial T}\left(c - \frac{x}{2}\right) x\,\Delta T_0 & \text{for } x > 0; \\[2mm] -\mu_0 J_c \frac{\partial J_c}{\partial T}\left(\frac{x}{2} - d\right) x\,\Delta T_0 & \text{for } x < 0, \end{cases} \tag{9-33}$$

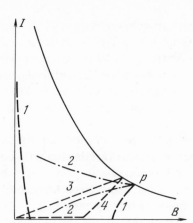

Fig. 9-6. Diagram defining the field and current range in which no flux jumps occur.

The coordinate x is counted from the line $E = 0$; the sections c and d are equal to the distance from this line to the surface of the layer under consideration. After integrating these relationships we find the mean value of the losses:

$$\overline{\Delta q} = \frac{\mu_0 J_c^2}{3 I_0} \frac{c^3 + d^3}{c + d} = \frac{\mu_0 J_c^2}{3 I_0} (c^2 - cd + d^2). \qquad (9\text{-}34)$$

Let us introduce the dimensionless quantity ι' representing the ratio of the total current in the conductor (transport current) to the maximum critical current:

$$\iota' = \frac{I}{2 b J_c} = \frac{c - d}{2b}. \qquad (9\text{-}35)$$

Transforming (9-34) with the aid of (9-35), we obtain

$$\overline{\Delta q} = \frac{\mu_0 J_c^2}{3 I_0} (1 + 3 \iota'^2) \, \Delta T_0. \qquad (9\text{-}36)$$

This equation differs from (9-9) by virtue of the factor $(1 + 3 \iota'^2)$ which enters into the new stability criterion

$$b J_{\text{cr}} \leqslant \left[\frac{3 c T_0}{\mu_0 (1 + 3 \iota'^2)} \right]^{\frac{1}{2}}. \qquad (9\text{-}37)$$

The maximum reduction in critical size (by a factor of 2) occurs for $\iota' = 1$. This result is explained by the fact that on reaching

the critical current the layer under consideration becomes equivalent to half the twice-as-wide layer not carrying the transport
current.

By using criterion (9-37) and existing data relating to the field
dependence of the critical current density for the particular material, we may plot a convenient diagram defining the range of current and field values in which no flux jumps should appear. The
general character of such a diagram appears in Fig. 9-6. The stability range is bounded by the broken line 1 obeying Eq. (9-37)
(with the equality sign) for a specified J_c (B) relationship. We
should remember that the T_0 in (9-37) also changes slightly with
changing magnetic field.

According to this equation the current density at the point at
which the boundary of the stability range intersects the horizontal
axis should be about twice as great as the current density at point
p, at which the critical current is stable.

Apart from stability in the range indicated, stability may occur in strong fields on reducing the external field below $H_{s.cr}$.
The role of thickness in (9-12) is here played by the depth of the
layer through which the field penetrates into the superconductor.
In calculating this quantity, strictly speaking, we should take account of the nonlinearity of the field distribution over the cross section, i.e., the specific J_c (B) relationship. However, from the practical point of view this region is of hardly any interest. In order
to illustrate the action of the "backing field," lines 3 and 4 have
also been plotted in Fig. 9-6; these characterize the increase in
the coil current (and hence the field in the center of the coil) which
occurs on simultaneously connecting all the sections (line 3) and
on connecting first the outer and then the inner sections (line 4).
If we neglect the diamagnetism of the winding, these lines should
be sections of straight lines. We see from the diagram how by the
aid of the "backing field" we may miss the "dangerous" region in
which flux jumps are able to develop.

The existence of a region of stability with respect to flux jumps
for sufficiently strong fields is already well known from many
experiments on the critical currents of short samples and small
test coils. The boundaries of the region in question move in the
direction of smaller fields (in agreement with the form of curve 1
in Fig. 9-6) as the current diminishes. As regards the accuracy
of quantitative estimates all the considerations set out in the pre-

vious section remain intact in this case. It should be remembered
that for sufficiently weak transport currents a transition into the
normal state may not occur even when flux jumps develop. This
will be considered in the next section.

9.5. Stability of the Transport
Current in Relation to Finite
Flux Jumps

The criteria which we have been considering determine the
stability of the superconductor with respect to the development of
flux jumps. When the dimensions of the superconductor fall below
the critical values given by Eq. (9-11) a flux jump should not develop. We see furthermore from the estimates of Section 9.3 that
the critical size is quite small; in weak fields, owing to the rise
in the critical current density, this permissible value may fall still
further. On the other hand, the complete exclusion of flux jumps
is, strictly speaking, an excessively rigorous requirement (although for certain special systems such as precision solenoids for
nuclear magnetic resonance it is highly desirable). In the majority
of superconducting magnetic systems, however, flux jumps by themselves, unless they convert the system into the normal state, produce hardly any other complications.

In this connection it is important to discover under what conditions a developing flux jump will lead to a specific new field distribution in the conductor without actually causing a transition into
the normal state; such a flux jump is sometimes called a partial
jump in contrast to "catastrophic." Concepts of partial jumps
were developed mainly by Hancox [98].

As the initial perturbation transforming into a flux jump develops, the parameters determining the velocity of the processes
involved alter. With increasing temperature the critical current
density diminishes, which causes a redistribution of the currents,
and the specific heat of the material also increases. At temperatures above approximately 4 K the main contribution to the specific
heat starts arising from the crystal lattice, and here the increment
in the specific heat obeys the well-known law $c \propto T^3$. Owing to
such a rapid increase in the specific heat, it may happen over a
certain temperature range that the intensity of the losses asso-

ciated with the rise in temperature will not suffice to increase the
heat content of the material, and any further rise in temperature
will then cease. The final result of the process, of course, will depend on the detailed volumetric loss distribution, the form of the
initial perturbation, and other factors. Under favorable conditions
the jump may die out over the entire volume, and as a result of
this under conditions of adiabatic insulation a new current distribution asserts itself, and the temperature assumes a constant value
corresponding to the losses created. Constancy of the resultant
temperature means that the current density at every point is incapable of exceeding a certain level corresponding to this finite
temperature.

We see from the foregoing description of the process that a
detailed calculation of the latter is far more difficult than an analysis of the stability of the isothermal distribution. Clearly the
averaging of the losses over the total cross section carried out in
the simplified derivation of the criterion (9-12) may here lead to
a far coarser divergence. Furthermore the study of even a one-
dimensional time process, and in particular the determination of
the ranges of parameters within which stabilization of the process
is feasible, constitute an extremely complicated problem, which
has never yet been completely solved.

We shall therefore confine attention to the simplified analysis
presented by Hancox [98]. Let us again consider a flat layer of
superconductor in which the field distribution has been established
by a simultaneous increase in the external field and the transport
current in the layer (Fig. 9-7a). The limiting possible case in
which all the magnetization currents vanish after the flux jump
and only the transport current remains corresponds to the field
distribution indicated by the broken line. The losses in such a
rearrangement of the field distribution could be calculated if we
knew the current (and hence field) distribution at every moment of
time. Hancox carried out his calculation on the assumption that
the thermal relaxation time inside the sample was much shorter
than the time required to establish the current distribution (for
rapid processes, as indicated earlier, the situation is precisely the
reverse). This assumption means that the heat is always able to
distribute itself over the thickness of the layer and the temperature in the layer is the same over the whole thickness at any given
instant. Since the field dependence of the current is not taken into

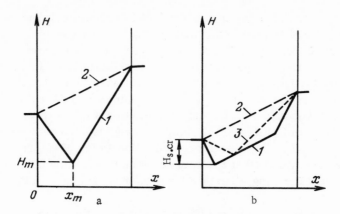

Fig. 9-7. Magnetic-field distribution in a flat layer of superconductor. 1, 2) Initial and ultimate field distribution; 3) distribution corresponding to current density J_3.

account, the absolute value of the current density at each moment
of time is also the same over the total cross section. The field
distribution at every moment has a characteristic "triangular"
form, and it is easy to show that x_m (the coordinate of the point m
at which $E = 0$ and $J = 0$) at any moment of time is equal to $b(1 - J_2/J)$
(here it is more convenient to reckon the coordinate from the
left-hand edge of the layer). Let us calculate the increment in the
heat liberated to the left of this point when the current distribution
changes by a small amount:

$$\Delta q_1 = \mu_0 \int_0^{x_m} d\xi\, J\,\Delta J \int_{x_m}^{\xi} \eta\, d\eta = -\frac{\mu_0}{2} \int_0^{x_m} d\xi\, J\,\Delta J (\xi^2 - x_m^2) =$$
$$= -\frac{\mu_0}{3} J\,\Delta J\, x_m^3 = -\frac{\mu_0 b^3}{3} J\left(1 - \frac{J_2}{J}\right)^3 \Delta J. \tag{9-38}$$

In an analogous way for the amount of heat liberated to the
right we obtain

$$\Delta q_r = -\frac{\mu_0 b^3}{3} J\left(1 + \frac{J_2}{J}\right)^3 \Delta J. \tag{9-39}$$

Summing (9-38) and (9-39) we obtain

$$\Delta Q = -\frac{\mu_0 b^3}{3} \int_{J_1}^{J_2} 2J\left(1 + 3\frac{J_2^2}{J^2}\right) dJ = \frac{\mu_0 b^2}{3}\left(J_1^2 - J_2^2 + 6J_2^2 \ln\frac{J_1}{J_2}\right). \tag{9-40}$$

In order to decide how correct our assumption is regarding the isothermal nature of the process we must estimate ΔQ in some other way. In order to make the problem as simple as possible, we may consider that a constant transport current $2bJ_2$ flows in the sample together with screening currents of density $-(J_2 + J_1)$ to the left of point m and $(J_1 - J_2)$ to the right. In the absence of an external field such a redistribution of the screening currents creates a field $-(J_2 + J_1)x$ to the left and $H_m + (J_1 - J_2)(x - x_m)$ to the right of point m. The energy associated with such a field distribution is given by the equation

$$E_n = \frac{\mu_0 b^3 J_1^2}{3}\left(1 - 2\frac{J_2^2}{J_1^2} + \frac{J_2^4}{J_1^4}\right). \tag{9-41}$$

For a rough estimate it is sufficient to assume that $\Delta Q \approx E_n$. The resultant equation describes a $\Delta Q(J_2/J_1)$ relationship differing from (9-40), although at the end points ($J_2 = 0$ and $J_2 = J_1$), of course, the two equations give identical results.

The heat which may be liberated in the superconductor during the foregoing redistribution of the field under conditions of adiabatic isolation is entirely expended in raising the temperature

$$\Delta Q = 2b \int_{T_1}^{T_2} c(T)\, dT. \tag{9-42}$$

If the temperature T_2 is higher than the value corresponding to the current J_2, it is impossible to reach an equilibrium state. Otherwise the process may attenuate for a certain intermediate field distribution. No resistance then arises relative to the transport current. We may thus describe the condition for the stability of the transport current relative to flux jumps in the following form:

$$\int_{T_1}^{T_2} c(T)\, dT > \frac{\mu_0 b^2 J_1^2}{6}\left(1 - \frac{J_2^2}{J_1^2} - 6\frac{J_2^2}{J_1^2}\ln\frac{J_2}{J_1}\right). \tag{9-43}$$

Using existing data regarding the temperature dependence of the critical current density and specific heat of the superconductor and with the aid of this equation we may construct the region of Fig. 9-6 in which flux jumps cannot lead the conductor into the normal state. The boundary to this region, shown by the upper dotted and dashed line 2, intersects the curve $J_c(B)$ at the point p, constituting the initial point of the region of "absolute stability" in which flux jumps never occur. This may be explained by the fact that when J_1 tends to J_2 the models used for the two criteria just considered become completely identical.

We shall not consider the behavior of the boundary to the new region of "limited stability" in any more detail. The point is that the resultant criterion is based on an extremely rigorous initial assumption and it should therefore usually lead to lower permissible current densities. This tendency is confirmed experimentally [93], despite the low accuracy of both the theoretical estimation itself and the computing data available.

The criterion (9-43) means that the superconductor transforms directly from the original to the normal state. But can such an original state in fact be realized? If an attempt is made to approach this state, the (9-12) type of criterion may be infringed far earlier, and then flux jumps will arise, leading to a completely different field distribution. It may also be that no transition into the normal state will take place during these jumps.

In attempting to calculate the sequence of such processes so as to determine when a "catastrophic" jump is liable to occur in the series of jumps a number of difficulties arise. The chief of these lies in the fact that even after a single jump the state of the conductor is somewhat indeterminate. The accuracy of the estimates is correspondingly lower in this case than it is when calculating a transition from an exactly specified state. The limiting stability condition was obtained by Wilson et al. [93] for two successive jumps, each of which completely "rubs out" the screening currents, leaving only the transport current $2bJ_2$. It is considered that in the interval between jumps the superconductor is able to cool completely to the original temperature T_1. The second jump develops when the field increases by an amount $H_{s.cr}$ and its distribution in the sample corresponds to Fig. 9-7b.

For reference purposes we here present the following equation, drawn from Wilson et al. [93], without giving its precise derivation:

$$\Delta Q = \frac{\mu_0 b^3}{3}\left(J_3^2 - J_2^2 + 6J_2^2 \ln\frac{J_3}{J_2}\right) +$$
$$+ \frac{\mu_0 H_2^3}{6b}\left\{\frac{2J_3 + J_2}{(J_3 + J_2)^2} + \frac{2J_3 - J_2}{(J_3 - J_2)^2} - \frac{2J_1 - J_2}{(J_1 + J_2)^2} - \frac{2J_1 - J_2}{(J_1 - J_2)^2}\right\}, \qquad (9\text{-}44)$$

where

$$H_2 = H_{\text{s.cr}} \left(1 - \frac{J_2}{J_1}\right); \quad J_3 = \frac{J_2 H_3}{\sqrt{H_3^2 + H_2^2} - H_2}; \\ H_3 = 2bJ_2.$$

(9-45)

The current density J_3 corresponds to the field distribution shown as a broken line in Fig. 9-7b (curve 3).

Using this equation we may calculate the position of the new boundary 2 to the region of "limited stability" in Fig. 9-6. This line also arises from point p and then usually corresponds to a monotonic relationship, so that the range of stability is widened in weaker fields. The curve based on (9-43), on the other hand, often has a positive derivative and sometimes intersects the axis I = 0. This means that the energies of the sceening currents alone are sufficient to heat the sample above the critical temperature in the particular magnetic field. It is reasonable to expect that in any real case the limit of stability will lie between two extreme positions.

Although the general characteristics of the behavior of the critical currents in different ranges of field may be qualitatively explained by using the various models here developed, remembering all that has so far been said on this score, quantitative agreement cannot be claimed.

Careful measurements of the critical currents in thin samples were made by Wilson et al. [93] using multicore conductors, and a comparison was drawn with the results of calculations based on the method just indicated; the effects of interactions between individual conductors appeared very strongly in these.

Effects of this kind in multicore conductors often have the result that in some respects these conductors become very similar to continuous (solid) superconductors of comparable cross section. Thus even if all the superconducting strands individually satisfy the stability criteria in question the conductor as a whole may be unstable. In the following chapters we shall consider the special characteristics of multiple-strand combined conductors.

Chapter 10

Multiple-Core Straight Conductors

10.1. Model of a Multiple-Core Straight Conductor

As already indicated, combined materials with internal stabilization are manufactured in the form of multiple-core conductors, since the production and use of a single-core superconducting
wire involves serious difficulties. The number of individual superconducting strands in the conductor may vary from 10-20 to several
hundreds or even thousands. In addition to this the combined conductor contains normal metal; this is either a pure metal with a
high electrical conductivity (usually copper) or an alloy with a
relatively low conductivity (such as a copper—nickel alloy of the
constantan type). Sometimes conductors containing all three components are made [99]. The combined conductor may also be a
cable twisted from individual wires consisting of superconducting
strands in a common sheath of normal metal.

In order to calculate the parameters of the combined conductor
it is convenient to construct a simplified model, more easily sus-
ceptible to theoretical analysis. The simplest and most convenient arrangement for model calculations is naturally a plane structure consisting of a large number of alternating layers of normal
metal and supercondcutor. For a correct representation of the
properties of a real conductor the thickness of individual layers
of the superconductor should correspond to the diameter of the
superconducting strands, and the total width of the model conductor
to the diameter of the combined conductor. Thus the thickness of

the normal layers in the model is uniquely determined by the specified occupation factor. Certain parameters (for example, the thermal resistance) of the layers do not correspond exactly to the same parameters in the real conductor. However, there is no difficulty in such cases in introducing certain effective quantities by varying the values of the parameters such as thermal conductivity, specific heat, and so forth.

The behavior of a combined conductor placed in a magnetic field and carrying a certain transport current is characterized by a number of important features which may duly appear in the properties of the plane model. As the simplest example let us consider the penetration of an external magnetic field into a semiinfinite space occupied by alternating layers of normal metal and superconductor when the field is parallel to the surface of the metal. We shall again study an idealized infinitely slow, i.e., isothermal process, in which any instantaneous field distribution is an equilibrium distribution. Clearly in this case all the currents induced in the normal metal are able to die away and in the normal regions the field will be uniform. Thus the presence of the normal metal will in no way affect the field distribution, which may now be constructed exactly as in the case of a single superconductor. The only difference is that the distribution will have a stepped appearance, since the field at the boundary of the superconductor retains the same value all through the normal zone (Fig. 10-1a). All distributions involving a transport current only (Fig. 10-1b) or an external field and a current (Fig. 10-1c) may be constructed in an analogous way.

If the combined conductor has an infinite length, then although the superconducting "layers" within it are not continuous the penetration of the field corresponds excellently in general features to the model under consideration. The field lines cannot penetrate too deeply through the breaks in the "layers," since they do not intersect all the superconducting wires forming the long layer. Thus the plane model should approximately reflect the properties of a circular combined conductor.

It may seem at first glance from the field distributions illustrated in Fig. 10-1 that on breaking a straight superconductor into small fractions the stability is not increased, and, moreover, that an index as important as the mean current density in the cross section diminishes. On comparing Figs. 10-1 and 9-5 it in fact

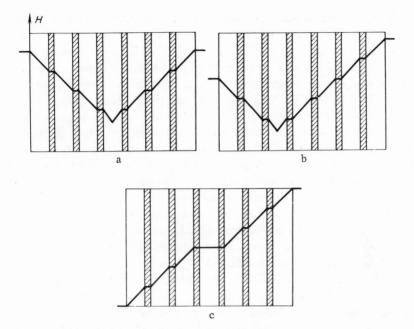

Fig. 10-1. Magnetic field distribution in the space occupied by alternating layers of normal metal and superconductor.

becomes clear that in all such cases the energy of the screening currents constituting the main source of instability in the conductor has comparable values. However, the inclusion of additional materials in the conductor may greatly alter the characteristics of the main processes, so that phenomena not playing any substantial part in the continuous conductor here become the decisive features.

10.2. Criteria of the Adiabatic Stability of a Straight Combined Conductor

If the material surrounding the superconductor is an alloy with a high electrical resistance and a poor thermal conductivity, any possible increment in stability will in fact here be negligible. This follows from a comparison between the conductor in question and a whole superconducting layer of equal size.

The development of a jump in the initial stages may as before be arbitrarily divided into small steps and the thermal balance calculated for each elementary step. The difference in the calculation lies in the fact that the decrease in the field across the thickness of the conductor corresponds not to the critical current density in the superconductor but to a certain effective current density, allowing for the finite degree of occupation of the cross section by the superconducor. Hence the total magnetic flux passing through the cross section in question increases. Since at the initial instant the currents in the shell are small, the increasing losses take place mostly in the superconductor. On the other hand, a certain proportion of the heat should be capable of being transmitted to the sheath. However, it is very difficult to take a precise account of these opposite effects.

We note that, if the material filling the space between the superconducting layers (strands) is a dielectric, there are hardly any qualitative differences in the development of the jump at the initial stage. A case of practical interest in which a model of this kind is realized is that of porous sintered Nb_3Sn, which was intensively studied in the initial stages of flux-jump investigations [100]. The increase in the permissible field $H_{s.cr}$ for such samples on passing through λ point of liquid helium should evidently be ascribed to an increase in the proportion of heat passing into the helium filling the pores of the sample, i.e., an increase in the effective heat capacity.

The straight combined conductor with an alloy matrix briefly considered here is by itself of only limited practical interest, since as we shall show it is easy to arrange for the penetration of the field to be effected in a more favorable manner in such a cable. Of considerably greater importance is the case in which the matrix of the conductor is made from material with a high electrical and thermal conductivity. In this case a decisive factor is the difference in the properties on which the rate of heat transfer and the rate of magnetic-field diffusion depend in such materials. Whereas for alloys the ratio of the field and temperature diffusion coefficients D_{em} and D_{te}, determined, in accordance with (9-6), by the equation

$$\frac{D_{em}}{D_{te}} = \frac{c\rho}{\lambda \mu_0},$$

(10-1)

is usually 10^3-10^4, for pure metals in the region of 4 K this ratio
is extremely small and may amount to 10^{-3}-10^{-4}. In other words
in pure metals thermal equilibrium is established before the magnetic field is able to redistribute itself in any marked fashion; in
contrast to superconducting alloys this conclusion remains valid
for any velocities of these processes.

Thus in a combined conductor with a pure copper matrix (even
for a relatively small proportion of copper) the rate of development of a flux jump embracing a considerable part of the cross
section falls sharply. In this case it may be considered that the
temperature of the matrix is the same over the total cross section.
The temperature inside the superconducting strands is certainly
higher than the temperature of the matrix. However, owing to the
protraction of the characteristic time of the jump, a far greater
proportion of the heat evolving in the superconductor will pass
into the matrix. Hence in calculating the energy balance for the
jump there are far better grounds for regarding the process as
isothermal than there were in the case of the continuous superconductor. For the same reason in all calculations it is essential
to allow for a certain average heat capacity of the conductor, including part of the heat capacity of the matrix, whereas for an
alloy matrix it would be more correct in general to consider simply the heat capacity of the superconductor itself.

The foregoing considerations enable us to extend all the criteria hitherto established to the case of combined conductors for
comparable configurations of field penetration. It is only necessary to introduce certain corrections into the corresponding equations, by averaging the effective values of the corresponding parameters relating to different components of the conductor. However,
no reliable method of calculating these corrections has been proposed. For high occupation factors of the cross section by the
superconducting material the correction factors clearly differ
little from unity. Since the accuracy of the original criterion (9-11)
was not very great we may rest content with this approximation.
We note that as in the case of a conductor with an alloy matrix the
field distribution over the cross section may be "assigned" a more
favorable form in the present case as well (see Section 11.2).

The adiabatic condition assumed in deriving the stability criteria means in the present case that heat transfer from the external boundary of the conductor to the surrounding is not taken into

account. Although for certain systems this assumption is excessively severe (we shall be considering some cases of this kind), it
should be remembered that in the overwhelming majority of magnetic systems based on conductors with internal stabilization the
windings are impregnated with various compounds having a low
thermal conductivity. Furthermore, if we accept the adiabatic condition we may properly assume that the flux jump embraces a
large number of neighboring turns.

Since a certain proportion of the heat evolved in the jump is
absorbed by the matrix material on account of its heat capacity,
the mechanism underlying the "stopping" of a partial jump by
reason of an increase in heat capacity may be more effective in
combined conductors in which the matrix material includes metals
with high specific heats. This principle [98] is the basis of the method of enthalpy (or adiabatic) stabilization (Section 3.5). The idea
of this method was proposed by Hancox, who calculated the energy
balance for a partial jump. This calculation was simplified by assuming that the temperature was the same at all points of the cross
section. Hancox correctly indicated the role of a copper matrix
in ensuring the isothermal condition, and successfully tested small
coils with a combined conductor containing up to 75 vol.% lead.
The critical current of the coil corresponded to the current of short
samples of superconductor, while the current density averaged
over the cross section was 1.25×10^4 A/cm^2. The choice of lead
as heat-absorbing material was due to its relatively high specific
heat (the Debye temperature Θ_D of lead is one of the lowest for
metals and equals approximately 95 K).

It would appear that the method of enthalpy stabilization should
have special advantages for certain specialized systems, such as
the class of so-called levitrons based on an annular turn carrying
a "frozen" current freely suspended in a vacuum chamber. The
use of pipes for introducing and removing helium in such systems
is fundamentally undesirable. The period of operation of the device is fixed by the heat capacity of the suspended magnetic system,
so that the heat capacity (specific heat) of the conductor in this
case plays a double role.

The technique of stabilization by means of substances with
a high specific heat has not received any special development.
Methods of calculating such conductors more refined than the original Hancox scheme have as yet been hardly developed; hence

this scheme is continually repeated in calculating more complicated twisted conductors.

10.3. Criterion of Electrodynamic
Stability for Combined Conductors

As already indicated, the assumption as to the complete adiabatic insulation of the turns in the conductor may be unduly severe. This situation appears most clearly in systems without any
impregnation, immersed in liquid helium. For plane conductors
wound in the form of individual disk coils, the use of impregnation
within the coil does not exclude the possibility of individual disk
coils being cooled from the ends.

Since the development of a flux jump inside the conductor in
the presence of a copper matrix may be considerably extended in
time, the heat evolved during the jump may be largely carried out
into the surrounding medium. In the limiting case we may regard
this mechanism, which prevents subsequent overheating, as fundamental, and the influence of the specific heat of the conductor material may be entirely neglected. It is clear that, subject to this assumption, the calculations will be executed with a certain reserve.

Thus for an adequate rate of heat outflow the heat generated
in the conductor will succeed in dissipating without any further
temperature rise. The current in the superconductor will then be
able to stabilize, i.e., any further jump development will cease.
This mechanism of flux-jump stabilization forms the basis of the
electrodynamic method of stabilizing combined conductors [101,
102]. Since the processes of heat outflow into the external medium that
underlie this mechanism are the same as those underlying cryogenic
stabilization, the two mechanisms have a great deal in common.
Furthermore, as we shall show, the stability criteria may almost
coincide in compatible situations. Thus, since the final intent of
stabilization methods is that of increasing the current density in
the windings, in subsequent calculations we shall seek to discover
what factors will produce a gain in current density when using
the method of electrodynamic stabilization.

Let us consider the development of a jump from this point of
view. The initial field distribution in the model chosen is determined by the effective current density in the cross section, equal
to $K_s J_c$ for distances exceeding the dimensions of the layers,

where K_s is the volumetric occupation factor for the superconductor. If as the result of an initial perturbation there is a temperature rise ΔT, the field distribution will start rearranging itself, tending toward a new distribution corresponding to a new current density $J_c + (\partial J_c/\partial T)\Delta T$.

During this rearrangement the basic currents will continue flowing through the superconductor, since in the superconductor the current cannot drop below the critical value. The electric field created during the motion of the lines of magnetic field inside the conductor may be regarded as uniform over the cross section. Hence the heat evolution of density JE will principally arise in the interior of the superconducting layers.

In estimating the average field E we should, generally speaking, allow for the influence of both components of the conductor. If we consider that the initial heating occurred instantaneously, then the field will penetrate into the normal metal in a time

$$\tau_n \approx \frac{\delta^2 \mu_0}{\rho_n}, \qquad (10\text{-}2)$$

where δ is the characteristic minimum size of the regions of normal metal. This time is of course much shorter than the time required to establish the field distribution over the entire conductor, which is determined by the second part of Eq. (9-6). Hence the current "excesses" $\Delta J_c K_s \times 2b$ are quickly redistributed between the superconductor and the normal metal "connected" in parallel, and the field will be determined by the equation

$$E = K_s \Delta J_c \left/ \left(\frac{K_s}{\rho_f} + \frac{1 - K_s}{\rho_n} \right) \right. = \theta_{\text{eff}} K_s \Delta J_c. \qquad (10\text{-}3)$$

Hence the average loss density referred to the whole cross section at the initial instant is

$$\dot{q} \approx \left[K_s^2 \left/ \left(\frac{K_s}{\rho_f} + \frac{1 - K_s}{\rho_n} \right) \right. \right] J_c \Delta J_c = K_s^2 \rho_{\text{eff}} J_c^2 \frac{\Delta T}{T_0}. \qquad (10\text{-}4)$$

Wilson et al. [93] obtained a similar expression (not allowing for ρ_f) distinguished by a coefficient $\pi^2/12 \approx 0.8$. The losses were estimated by dividing the increment in the total magnetic energy of the screening currents by the characteristic attenuation time of the ground harmonic in the magnetic field distribution. Equation (10-4) is more accurate, since the contribution of the higher harmonics, which die out far more rapidly, is quite considerable at the initial instant. This at first glance quite unimportant refinement of the numerical factor nevertheless plays an important part subsequently when comparing with the criterion of cryostatic stabilization (4-22).

Equation (10-4) also describes the mean density of heat evolution in a combined conductor in which the current was initially critical, after which its temperature increased by ΔT. The reason for this coincidence is perfectly clear. The current density in the superconductor is everywhere equal to the critical value when the field penetrates to the middle of the conductor, and in the initial stages of the process this situation remains approximately intact.

The applicability of the expression for the density of the losses to the problem already considered enables us to pursue further investigations along similar lines. If the basic thermal resistance is associated with heat transfer to the liquid helium, then in complete conformity with the stability analysis presented in Section 4.5 we obtain an analogous stability criterion

$$A\dot{q} \leqslant hP\Delta T$$

or

$$\frac{K_s^2 J_c^2 A^2}{hPAT_0} \left/ \left(\frac{K_s}{\rho_f} + \frac{1-K_s}{\rho_n} \right) \right. = \frac{J_{c^0}^2 \, \mathrm{eff}}{hPA\,(T_c - T_t)} \leqslant 1. \qquad (10\text{-}5)$$

This relationship naturally coincides with the stability criterion $\alpha \leq 1$ for cryostatic stabilization, when it is usually assumed that $\rho_f \gg \rho_n$ and $K_s \ll 1$.

Let us consider the possibility of increasing the mean current density in conductors with electrodynamic stabilization. It would appear that the satisfaction of criterion (10-5) should automatically ensure the possibility of working under conditions of cryostatic stabilization, even for the maximum critical current of the conductor. However, the difference between the presentation of the problem of stabilization in the two cases under consideration may have the effect that, despite their formal identity, the two criteria will give markedly differing numerical values for the admissible parameters of the conductor. The point is that, owing to the essentially nonlinear temperature dependence of the two most important quantities entering into Eq. (10-5), namely, the heat-transfer coefficient h and the resistivity of the superconductor in the resistive state ρ_f, the choice of their numerical values depends on the permissible initial degree of superheating for which the conductor is designed.

In cryostatic stabilization a quantity of the order of 1 K is usually taken as an acceptable level of superheating (often more than this in large systems) so as to ensure that the conductor may always remain quite stable on subjection to very severe perturbations and steady heating due to the constant transport current. In the case under consideration the quantity ΔT may be determined for the optimum values of the parameters in Eq. (10-5). Actually the satisfaction of such a relationship at any instant means that subsequent heat evolution stops altogether (we remember again that no allowance is made for the heat capacity of the conductor). Thus in executing calculations based on Eq. (10-5) it is perfectly reasonable to take maximum values up to 1 $W/(cm^2-K)$, corresponding to superheatings of the order of 0.2 K, for the coefficient h. For such superheatings the influence of the resistance ρ_f may clearly be neglected. In this case there may be a gain of about two or three times in the current density as compared with cryostatic stabilization, in which no reserve of stability would be obtained at all for the maximum value of h.

On reducing ΔT the heat-transfer coefficient diminishes, but the coefficient ρ_f tends to zero, and in this limiting case the problem becomes insufficiently determined, since the intensity of the losses also tends to zero; thus stability is guaranteed. A reasonable presentation of the problem on a theoretical basis therefore requires that the parameters of those perturbations against which the particular stabilization mechanism is intended to "work" should be fairly sharply defined. It is clear that without the accumulation of adequate experimental material as to the character of such actions the further development of the problem in this direction may prove unfruitful. However, the derivation of any information as to the character of the interaction between the turns and the parameters of the fundamental processes in such windings from experimental data is extremely difficult.

On the other hand, we note that the foregoing analysis is related to the stability of a static field configuration. In actual practice it is always important to know what are the permissible limits
in the rate of change of the external conditions (i.e., the field, the
current in the winding, and so on). In ensuring adiabatic stability
this question had no decisive significance, since it is considered,
for example, in the case of the Hancox partial jumps, that even the
whole magnetization energy is unable to heat the conductor above
the permissible limit. For a conductor only calculated with respect to dynamic stability, on the other hand, the rate of change of
the field is determined by the possible rate of outflow of the evolving heat. The permissible velocity should in any case be such that
the field distribution in the conductor should correspond to that
which we have chosen, i.e., that it should satisfy the condition

$$\tau_0 \gg \tau_{em} \approx \frac{\mu_0 b^2}{\rho_n} (1 - K_s), \tag{10-6}$$

where τ_0 is the characteristic time of change in the external parameters.

As characteristic resistance we here use ρ_n since we are
considering a limitingly rapid process in which this resistance may
become extremely large. On satisfying the original criterion (10-
5) with $\rho_{eff} = \rho_n$ and also condition (10-6), the losses (heat) evolved
will also succeed in dissipating (we showed earlier that $q \approx E/
\tau_{em}$).

As in the case of cryostatic stabilization, for certain configurations of the conductors the decisive factor from the point of view
of heat release may be the thermal conductivity of the metals forming the combined conductor. Thus, if the configuration of the
conductor approximately corresponds to the model under consideration (combined strip-like conductor), then, depending on the particular geometrical relationships, the principal thermal resistance
may arise either in the interior of the superconducting layers (assuming these to be thick enough) or in the interior of the normal
layers (for external heat transfer from the ends). The study of
dynamic stability for these cases is more complicated and should
again amount to the analysis of an equation in partial derivatives,
since the thermal resistance here is not concentrated. The in-

fluence of the thermal resistance in the interior of a superconductor was studied in fair detail by Hart [102], who also calculated the case in which the external thermal resistance was comparable with the resistance inside the superconductor. However, in these problems only a continuous superconducting layer was studied, with artificial assumptions regarding the small value of the electrical resistance of the superconductor. Thus it was assumed in the model that the normal metal did not take any part in retarding the flux jump. We shall therefore now reproduce the results of Wilson et al. [93], which correspond to a model similar to our present one. The parameters taken in this model may easily be related to the parameters of a specific conductor.

We shall consider that the external thermal resistance is small, while the thermal conductivity of the copper matrix is large enough for the temperature at all the boundaries of the superconductor to be equal to the temperature of the surrounding medium. If by way of simplification we assume that the initial perturbation leads to a temperature rise ΔT in all the superconducting layers, we again obtain Eq. (10-4) for the power of the heat losses; this should now be referred to the total cross section of the layers of the superconductor

$$q \approx \frac{\rho_{\mathrm{eff}} K_s}{T_0} J_c^2 \Delta T. \qquad (10\text{-}7)$$

On the other hand, for uniform heating of the layer with a specified specific power P, a parabolic temperature distribution will occur

$$\Delta T = -\frac{P}{2\lambda_s}(x^2 - d^2), \qquad (10\text{-}8)$$

where d is the half width of the layer of superconductor. The average value of ΔT over the layer is determined by the equation

$$\overline{\Delta T} = \frac{1}{3\lambda_s} P d^2. \qquad (10\text{-}9)$$

Hence the thermal resistance of the superconductor will in order of magnitude be

$$\frac{\overline{\Delta T}}{2Pd} = \frac{d}{6\lambda_s}. \qquad (10\text{-}10)$$

In other words, for an average superheating of $\overline{\Delta T}$ the heat flow from the layer will be 2dP. The stability requirement may again be formulated as the condition that the intensity of heat outflow should exceed the heat evolution for the particular initial overheating* ΔT

$$\frac{\rho_n K_s}{(1-K_s)T_0} J_c^2 < \frac{3\lambda_s}{d^2}. \tag{10-11}$$

It is not hard to verify that at the corresponding limit this equation also transforms into Eq. (4-68), which we obtained when studying thermal resistance (to a factor of 1/2). We remember that Eq. (4-68) corresponded to the appearance of a negative derivative in the temperature dependence of the resistance of unit length of conductor, i.e., to instability.

Criterion (10-11) determines the upper permissible limit for the thickness of the superconducting strands or the plane layer in the conductor (2d). Numerical estimates based on this criterion for a relatively high occupation factor $K_s = 0.5$ (characteristic of conductors with internal stabilization) give a value of the order of 40 μ for the thickness d [93]. Satisfaction of the criterion (10-11) does not of course mean that the conductor will actually be stable as a result of dynamic stabilization. It is also essential that all other thermal resistances "connected" in the path of the heat flow be fairly small.

Let us now find the criterion of dynamic stability for the conditions in which the decisive thermal resistance is that of the normal metal. A case of practical importance from this point of view is that of a thin, wide strip cooled from the ends. In order to estimate the thermal resistance we may use an equation of the (10-10) type, replacing d by the strip half width a:

$$\frac{\overline{\Delta T}}{2Pa} \approx \frac{a}{6\lambda_n}. \tag{10-12}$$

If the intensity of the losses is referred to unit volume of the copper strip we obtain

$$\frac{\rho_{\text{eff}} K_s^2}{(1-K_s)^2 T_0} J_c^2 \leqslant \frac{3\lambda_n}{a^2}. \tag{10-13}$$

---

*A similar result was obtained by Chester [45] in 1967.

This equation again simply determines the maximum permissible limit for the strip width. In view of the indeterminacy already indicated as regards the choice of numerical values of the parameter ρ_f in the criteria (10-5), (10-11), and (10-13), for any specific estimates we should use all these equations with exactly the same value of the resistance ρ_{eff}.

In principle we may encounter a case in which all three ways of releasing heat from the superconductor are characterized by approximately equal thermal resistances. Following the principles of the foregoing calculation, in this case we should have to derive criteria containing the sum of the corresponding thermal resistances. However, such an equation would not have any practical value since each of the terms would be calculated with an inadequate accuracy. Furthermore it follows [102] that the calculation would not amount to simple summation. We note, furthermore, that by itself the approximate equality of all the thermal resistances may signify that the combined conductor is not really constructed in the best possible way, and that it is difficult to achieve any marked increase in the current density of the coil as compared with the case of cryostatic stabilization.

The model of the combined conductor under consideration is entirely applicable for estimating the properties of real multicore conductors with a copper matrix. Certain difficulties in the specific calculation of thermal resistance may arise if the superconducting strands lie very close together, when the thermal resistance of the copper may become quite substantial. It should be remembered that the use of multicore conductors without impregnation to prevent their mechanical displacement involves serious difficulties. If, however, the winding is impregnated with dense compounds, one does not have to rely on the action of the dynamic-stabilization mechanism in pure form.

The correspondence of the model under discussion to any strip-like conductor situated in a field parallel to the surface is quite obvious, and the relationship between their parameters is perfectly clear. Strip conductors usually contain one or two layers of superconducting material of extremely small thickness, as determined by the requirements of mechanical bending strength. For two-layered materials the thickness of the intermediate layer of normal metal is also small. For these reasons it is easy to satisfy the criteria of adiabatic stability for such conductors.

However, in any magnetic system there are parts of the winding in which the field has considerable components normal to the plane of the strip. This configuration leads to certain singularities

in the calculations. For simplicity we shall consider a disk coil
in which successive turns of the strip are laid one over the other
along the radius of the coil. We also assume that the total transport current in the strip is small compared with the maximum
critical current, and we shall only consider the influence of the
currents screening the normal field components, not considering
their interaction with the screening currents for the field $H_{\parallel}$ lying
in the plane of the strip. This may mean, for example, that at first
the longitudinal field $H_{\parallel}$ penetrated completely into the superconductor as the result of a partial flux jump, after which the normal
field $H_{\perp}$ was "connected." Such a simplification may be rather
excessive, and the results obtained will bear only a qualitative
character.

It is easy to see that the pile of strips in a field perpendicular
to the plane of the strips will behave in the same way as a continuous layer of superconductor with a thickness equal to the strip
width $2a$ (the end effects, which are considerable for the extreme
turns, are not taken into account). The picture of field penetration
will correspond to an average current density of $J_c K_s$. This enables us to use all the equations already derived for the energy
density. Furthermore it may easily be shown that even Eq. (10-4),
defining the density of the losses, retains its form (but only for the
parts of the strip to which the field $H_{\perp}$ has penetrated). In fact the
field E arising during the change in the magnetic field $H_{\perp}$ lies in
the plane of the strip and therefore has the same value in the normal and superconducting regions. The currents excited by this
field run parallel and the corresponding conductivities add, i.e.,

$$\rho_{eff} \approx 1 \left/ \left(\frac{K_s}{\rho_s} + \frac{K_n}{\rho_n} \right) \right. . \tag{10-14}$$

We note that this conclusion does not depend in any way on
whether the two components of the winding are in electrical contact.
If, however, the cooling of the superconducting strip takes place
without the participation of the normal strip (for example, if superfluid helium penetrates into the gaps between the layers), then the
two strips may be completely isolated from one another (in sharp
contrast to the requirements of cryostatic stabilization).

Thus in the approximation under consideration we may use
Eqs. (10-4), (10-11), and (10-13) in order to estimate the stability
of the conductor for the parameters chosen. It is only required

that in (10-13) the half width of the strip should be replaced by the distance $H_\perp/J_c$ through which the field $H_\perp$ has penetrated into the interior of the disk coil (in the central regions of the conductor there is no heat evolution, and for the heat-transfer mechanism under consideration these regions are also not important). As a result we obtain

$$H_\perp^2 \leqslant \frac{3(1-K_s)^2 T_0 \lambda_n}{\varrho_{\text{eff}} K_s^2}. \tag{10-15}$$

In an analogous way in Eq. (10-5) A should mean only that part of the total cross section of the conductor to which the field $H_\perp$ has penetrated, while P should be taken to mean the whole perimeter of the conductor accessible for cooling. If, however, the superconductor transfers heat directly to the liquid helium, P will again equal $H_\perp/J_c$.

Thus if we have a favorable field distribution over the cross section of the conductor (the screening currents occupying only part of the cross section), and use the whole perimeter for cooling, the condition of dynamic stability (10-5) may be satisfied for a more advantageous occupation factor K_s. In practice this simply applies to the case of a strip-wound disk coil; for other configurations the screening currents embrace the whole cross section of the straight conductor.

Chapter 11

Twisted and Coiled Combined Conductors

11.1. Penetration of a Magnetic Field into a Coiled Combined Conductor

The model of a straight, infinitely long conductor analyzed in the preceding sections was rather a convenient abstraction. Let us now consider some of the special features characterizing this model in more detail.

As already noted, the instability of the conductor arises from unattenuated currents circulating in the superconducting strands and screening the inner regions of the conductor from the penetration of the magnetic field. Thus if the magnetic field could be caused to penetrate into the inner regions in some way this source of instability would be eliminated. The superconductor would then form a set of independent filaments and the pattern of field penetration would correspond to Fig. 9-3. Thus, if only each individual filament were stable (and this is an incomparably less severe requirement than the stability of the entire conductor) the conductor as a whole would also be stable.

Since a real conductor always has a limited length, it is natural to consider whether the magnetic field might be able to penetrate into the inner regions missing the superconducting barriers, i.e., from the ends of the conductor. The inner layers of the normal metal are indeed "open" at the ends for the penetration of the magnetic field, which should "diffuse" from there to the center of the conductor. For the model of the infinite layer the characteristic time of such a process (i.e., the time τ_0 within which the field penetrates substantially into the conductor) may be estimated very

299

simply if we neglect the slight diffusion of the field inside the superconductor. In this approximation the problem will be one-dimensional, since the changes taking place in all the quantities will only
proceed along the conductor. As for all processes described by
the diffusion equation (9-5) we shall have

$$\tau_0 \approx \mu_0 l^2 / \rho_n, \tag{11-1}$$

where l is the half-length of the conductor.

Let us estimate this time for a conductor 20 m long with a
pure copper matrix. The resistivity of copper at 4 K may be taken
as 2×10^{-10} Ω-m. Hence

$$\tau_0 \approx \frac{10^2 \times 4\pi \times 10^{-7}}{2 \times 10^{-10}} \approx 6 \times 10^5 \text{ sec,} \tag{11-2}$$

i.e., approximately 170 h. This value may appear very unlikely,
but direct measurements [93] give values of several minutes for
a conductor 30 cm long and about 240 h for a length of 22 m, which
agrees with our simple estimates.

If the matrix of the conductor is made from hard alloys, which
may have a resistance several thousand times greater than that
of copper, the penetration time will be as many times shorter; however, it is clear that any straight conductor of real length may in
all respects be regarded as infinite. The penetration of the field
into the conductor with finite cross-sectional dimensions having
individual superconducting strands instead of the plates may of
course be more complicated; however, this will only affect the
numerical value of the coefficient in Eq. (11-1).

It may nevertheless be shown that, if we twist the whole conductor along its axis, the penetration of the field into the same
conductor may be accelerated significantly. In order to carry out
a simple analysis of the penetration of the field into a twisted conductor it may be more convenient to consider the "reverse" process. Let us imagine that the field has penetrated into the normal
metal along the entire length of the conductor. This may easily be
done if we heat the conductor above the critical temperature. The
source of the external field is then disconnected. The conductor
then retains a certain magnetization created by loop currents which
will flow through the superconducting strands along the conductor,

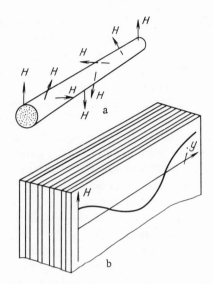

Fig. 11-1. Magnetic-field distribution in a round conductor (a) and in a plane model (b).

being closed near the ends by the normal metal of the matrix. The inclusion of a certain finite resistance into the current circuit in the end regions leads to the attenuation of the current.

It is not difficult to see that for this presentation of the problem the current distribution will correspond entirely to the original picture, and the characteristic times of the processes will therefore coincide. Let us imagine that the conductor is straight but the initially connected field is "twisted" (Fig. 11-1a). Such a field configuration may be created, for example, by means of two conductors with currents of different sign wound on a cylinder coaxial with the conductor under consideration. After disconnection of the external "twisted" field, the central conductor retains its original magnetization; the direction of this magnetization changes (rotates) on passing along the axis of the conductor. We may now study the attenuation of one component in the initial distribution of the magnetization vector by itself, for example, the x component, since on disregarding insubstantial nonlinear effects the two components attenuate independently in the axisymmetrical conductor. The initial distribution of the x component of magnetization is now given by a sinusoidal function.

The problem is thus clearly reduced to the original form; only the role of the total length of the conductor is now played by the half wavelength of the sine wave (Fig. 11-1b). For calculating the time of field penetration we may once again use Eq. (11-1), replacing l by a quarter of the total twist pitch (i.e., a quarter of the distance over which a particular superconducting strand executes one rotation around the axis of the conductor). The time τ_0 may be made quite short; even for a pitch of 2 cm it amounts to approximately 0.1 sec (in the case of a copper matrix).

The time of field penetration into the conductor τ_0 enables us to separate all the processes associated with changes in the external field into two categories. If the characteristic time of the process (for example, the time during which the external field increases) is much shorter than τ_0, then the behavior of the conductor will correspond to the properties of a straight conductor, as considered in the preceding section. However, if the process takes place during a time much longer than τ_0, then as far as the external field is concerned the conductor resembles a set of independent superconducting strands. Thus, for example, if the field increases rapidly, flux jumps may be created in the sample, even if individual strands are stable with respect to jumps, whereas the conductor as a whole fails to satisfy the corresponding criteria.

The very slow penetration of the field into a straight, untwisted conductor may in certain cases nevertheless lead to complications. The diamagnetic moments of the turns create a certain additional field around the winding, adding on to the field of the transport current. The resultant field in the center of a solenoid with a winding of rectangular cross section then increases slightly, while the degree of homogeneity of the field in the central region will also increase. This effect may easily be understood if we picture a winding based on an ideal diamagnetic which completely "squeezes out" the field. The penetration of the field through the inner regions of the turns means that the diamagnetic moment of the turns diminishes. Thus as time progresses the field in the center of the system will also decrease slightly, and the homogeneity of the field will worsen. Thus this effect may have a certain significance in those systems in which an exclusive (extreme) stability of the field or an exact knowledge of its exact distribution in the working volume is required, for example, in precision systems for nuclear magnetic resonance and in systems for bubble chambers.

In a real magnetic system the rate of field penetration into a conductor not twisted during actual manufacture will, inter alia, be determined by such poorly reproducible factors as random twisting during the winding process (for circular conductors), the existence of joints between individual sections of conductor, the possibility of local overheating (for systems with cryostatic stabilization), and so on. Thus it is not so easy to calculate the initial field distribution and the course of the process reasonably accurately. Similar difficulties arose, for example, in making the magnetic system for the bubble chamber of the European Center for Nuclear Research. In order

to monitor the magnetic field distribution, which had to be known in order to make
a precise calculation of the trajectory of the test particles, a vast number of Hall sen-
sors had to be placed in the working space of the chamber.

As already indicated, the use of twisted conductors enables
us in relatively simple ways to achieve a more advantageous distribution of the external field in the conductor, which reduces the
undesirable energy of the screening currents. However, as we
shall subsequently see, the associated possible gain in current
density is not very great in the case of enthalpy stabilization, and
in the case of electrodynamic stabilization it is completely absent.
In this section we have not considered the existence of a transport
current in the conductor; the distribution of this current over the
cross section does not in fact differ in any radical manner for
twisted and straight conductors.

11.2. Distribution of Transport Current
in Twisted Conductors and the Stability
of the Latter

Let us now consider the character of the field distribution in
a twisted conductor when the current rises after applying an external uniform field. We shall consider that sufficient time has elapsed
after the application of the field for the field to have penetrated completely into the normal regions of the conductor (continuous line in Fig. 11-2). First of all the current may occupy only
the very outermost layers of the superconducting strands (broken
lines). The field lines of the transport current are in fact concentrated around the axis of the conductor and cannot therefore intersect the outer layer of the superconductor, until the current in
this layer exceeds the critical value. The same situation is repeated with the second, third, and subsequent layers (it is assumed
that the number of layers is not too small). It is an important
feature that this picture of penetration is independent of whether
the particular conductor is twisted or not, provided that the position
of the individual strands in a particular layer does not alter along
the length of the conductor.

When the external field and the current in the conductor increase simultaneously or are sufficiently slow, the final field dis-

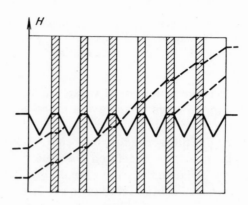

Fig. 11-2. Magnetic field distribution in a twisted
conductor.

tribution will always correspond to Fig. 11-2, provided that the
distribution is always established in an equilibrium manner, i.e.,
without flux jumps embracing substantial regions of the cross section. To this end the characteristic time for the increase in the
external field should greatly exceed the quantity τ_0 determined by
the twist pitch.

Let us consider how the character of the field distribution in
the twisted conductor influences its stability. We shall first of all
consider that the current in the conductor is increased to the maximum critical value. We thus have to study the stability of the distribution corresponding to the very last moment before the cross
section is completely filled with the critical current. It is not difficult to see that from the point of view of dynamic stability the
conductor under consideration is no different from a straight conductor, since the mean current density determining the initial power of heat evolution is as before equal to $K_s J_c$ in all regions.

A slightly different position arises on using the criteria of adiabatic stability. Although the final field distribution is here identical with the distribution of the straight conductor, the initial distribution is equivalent to the distribution in two layers of half thickness. Thus, other conditions being equal, the simple twisting of the
conductor is equivalent to the use of a conductor of half the diameter, which enables us to increase the current density by approximately a factor of 2 while satisfying all the stability criteria.

It should be noted that this kind of penetration of the transport-current field into the combined conductor strictly corresponds to the case of a single conductor only. In a fairly dense winding, owing to the action of the magnetic fields of the neighboring turns, the penetration picture may be more favorable, since the field lines of the transport currents in individual turns do not necessarily embrace a particular turn but may pass along the whole layer. Of course a field with this configuration penetrates into the twisted superconductor in the same way as a uniform external field. However, no accurate estimates have yet been made for the influence of this "proximity effect."

If the parameters of the conductor do not satisfy the stability criteria for the maximum critical current, we must calculate the permissible current values for the corresponding stabilization mechanism. In the case of dynamic stabilization once again there are no differences relative to the case of the straight conductor, since independently of the form of field distribution over the cross section all the strands of the superconductor are in the critical state, i.e., the current density in these strands is equal in absolute magnitude to J_c. The relationship for the losses at the initial instant of the jump determining the stability therefore retains its earlier form (10-4). As regards the problem of determining the stability limit with respect to jumps from the adiabatic criterion, it is not difficult to see that its solution is determined by Eq. (9-16). It is in fact clear simply from the nature of the field distribution (Fig. 11-2) that stability is ensured if the difference in the field strengths on the two sides of the conductor is no greater than $2H_{s.cr}$, i.e.,

$$\Delta H = \frac{I}{2b} < 2H_{s.cr} . \qquad (11-3)$$

The field $H_{s.cr}$ should of course be calculated from the mean density of the current in the cross section.

Slight differences relative to the problems already considered arise on calculating the stability of the transport current with respect to partial jumps. In this case incomplete penetration of the field occurs. Since in actual practice this situation arises only rarely in single conductors, we have not hitherto given it any special consideration.

As in the case of a single conductor, the most unfavorable situation corresponds to a transition directly from the initial field distribution (continuous line in Fig. 11-3a) to the final distribution

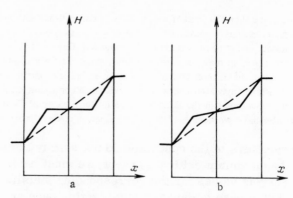

Fig. 11-3. Initial and final (broken line) magnetic-field distribution in a twisted conductor.

in which the transport current occupies the whole cross section uniformly (broken line), since the change in the magnetic energy is then greatest. Let us calculate the heat evolution for such a transition, once again making use of the Hancox assumption as to the uniformity of the temperature distribution over the cross section during the jump. The coordinate (counting from the middle of the sample) x_0 of the point at which the current density vanishes is determined by the equation

$$x_0 = b \left(1 - \frac{J_2}{J} \right). \tag{11-4}$$

The increment in the amount of heat which arises on changing the current density by ΔJ equals

$$\Delta q = \int_{x_0}^{b} JE \, dx \, dt \approx - \int_{x_0}^{b} J \int_{x_0}^{x} \Delta B \, d\xi \, dx =$$

$$= - \frac{\mu_0}{2} \int_{x_0}^{b} J \Delta J \, [(b - \xi)^2 - (b - x_0)^2] d\xi = \frac{\mu_0 J \Delta J \, (b - x_0)^3}{3}. \tag{11-5}$$

Integrating, we obtain

$$q = \int_{J_1}^{J_2} \mu_0 b^3 \, \frac{J_2^3}{3J^2} \, dJ = \frac{\mu_0 b^3 J_2^2}{3} \left(1 - \frac{J_2}{J_1} \right). \tag{11-6}$$

Finally,

$$\frac{\mu_0 b^2 J_2^2}{3}\left(1-\frac{J_2}{J_1}\right)\leqslant\int\limits_1^2 c\,dT. \tag{11-7}$$

We note that by virtue of the nature of the calculation we have here used the current density averaged over the cross section.

As before a less rigorous criterion is obtained for the case in which a flux jump occurs as soon as the field distribution becomes unstable. We shall here present (without detailed derivation) the equation obtained by Wilson et al. [93] on the assumption that the previous flux jump led to the complete occupation of the cross section by the transport current; the value of this current corresponds to the slope of the field-distribution line in the middle of the sample (Fig. 11-3b):

$$\frac{H_2^3}{6b}\left[\frac{2J_2-J_3}{(J_2-J_3)^2}-\frac{2J_1-J_3}{(J_1-J_3)^2}\right]\leqslant\int\limits_1^2 c\,dT; \tag{11-8}$$

here

$$J_3=\frac{J_2 b-H_{\text{s.cr}}}{J_1 b-H_{\text{s.cr}}}\,; \tag{11-9}$$

$$H_2=H_{\text{s.cr}}\left(1-\frac{J_3}{J_1}\right). \tag{11-10}$$

The partial occupation of the cross section of the twisted conductor with transport current means that some of the superconducting strands may not be used at all for current transport. It is nevertheless clear that in any particular winding a conductor of specific type will always occupy a certain finite region (and often the entire winding), and it is therefore quite impossible to achieve a situation in which the current will everywhere correspond to the maximum critical value for a particular local field.

In order to secure some idea as to the potentialities of twisted conductors in various fields it is useful once again to express in $B-I$ coordinates the boundaries to the regions of transport-current stability based on the corresponding adiabatic criteria, as in the earlier case of single conductors (Fig. 11-4, curves 1). The same

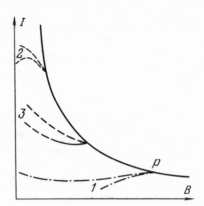

Fig. 11-4. Stability regions of the transport current in twisted conductors.

figure shows the stability boundaries for individual strands of the superconductor (curves 2) and a conductor not subjected to twisting (curves 3).

At the point p at which the corresponding criterion starts being satisfied, for a transport current equal to the maximum critical value the boundaries of the stability region merge with the curve expressing the field dependence of the critical current. At this point the current in the twisted conductor is more than twice as great as the current in a straight conductor, since T_0 decreases on increasing the field. This same effect of course influences the ratio of the currents for the curves determining the stability of individual strands of the superconductor and the straight conductor as a whole.

Experimental investigations [93] undertaken in order to achieve a detailed confirmation of the concepts of adiabatic stability here outlined in general agree closely with these; the quantitative agreement frequently obtained may be regarded as extremely satisfactory when we remember the large number of simplifications which had to be made during the calculations.

The effect of twisting the conductor may be demonstrated especially clearly by certain experiments in which the external field and the current in the sample are raised separately and in different orders. In the twisted sample (Fig. 11-5a) the critical currents obtained for a fixed current in the sample and a rising field (indicated by points) are on the whole even slightly higher than the values obtained for the reverse sequence of current and field connection (crosses). In

exactly the same conductor not subjected to twisting (Fig. 11-5b)
the current values suffered catastrophic "degradation" for the
first method, while for the case in which the external field was
established first there was hardly any marked difference between
the conductors.

Figures 11-5a,b also show the stability limits of the
conductor calculated theoretically by reference to various criteria.
The fine broken line represents the stability of individual strands,
the dotted and dashed lines correspond to the stability limits of the
conductor with respect to the field of the transport current, and
finally the broken lines in Fig. 11-5b represent the stability limit
of the conductor as a whole. The continuous lines represent the
critical current as a function of the external field. We see that the
agreement between the experimental data and the theoretical estimates is completely satisfactory.

Thus the proposed adiabatic stability criteria constitute a
reasonably reliable basis for estimating the properties of multicore combined conductors. As regards the mechanism of electrodynamic stabilization, although this influences the characteristics
of the combined conductors to some extent, its effects have not yet
been studied in pure form in any great detail. Since the conditions
of sample cooling are of decisive significance for this mechanism,
such investigations should preferably be carried out with reason-

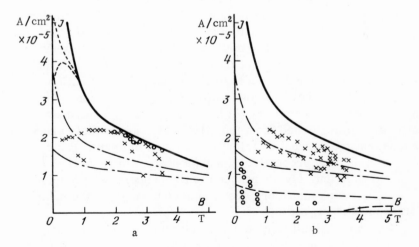

Fig. 11-5. Critical current density for twisted (a) and untwisted (b) samples.

ably large coils, for which the conditions of heat transfer may be
regarded as completely uniform.

We note that the twisting of the conductor as already indicated
has no effect on the efficiency of the electrodynamic stabilization
mechanism. The use of twisted conductors specially designed for
dynamic stability is only justified in those systems in which it is
desirable to ensure high stability of the field distribution in the
working volume (see Section 11.1).

11.3. Plaited (Transposed)

Combined Conductors

As already indicated, the nonuniform manner in which the current fills the cross section prevents the potentialities of individual
superfine conductors from being completely realized even in twisted multicore conductors. The "reserves" of magnetic energy so
created, always ready to be released when the current distribution
is equalized, as before constitute a source of instability in combined conductors of this kind. Methods of equalizing the current
distribution over the cross section of multicore cables have for
a long time been known in electrical technology. This problem
arises when alternating currents of fairly high frequency have to
be transmitted along such cables. In the case of high frequencies
the inner layers of the cable are screened by its outer layers in
the same way as in a superconducting cable under dc conditions.

For all the superconducting strands to be uniformly occupied
by current it is essential to make a conductor completely symmetrical with respect to each of the strands composing it. This means
that each strand should successively occupy all the positions characteristic of the other strands in different cross sections. In the
case of complete symmetry, in fact, the fact that no strands are
specially distinguished from the others as regards their properties
means that the current distribution remains uniform. It may also
be shown that the properties of such a system in respect of the
external field entirely correspond to the properties of a twisted
conductor. Such multicore systems have been given the name
"transposed."

It is not difficult to give a very simple example of a completely
transposed system. If in a multicore conductor all the superconducting strands lie at the same distance from the center in a single

circular layer, the desired symmetry with respect to the transport
current is clearly maintained. In order to ensure the symmetry of
such a conductor with respect to the external field it is sufficient
to carry out twisting in the longitudinal direction. However, although the preparation of such conductors is entirely feasible when
the diameter of the individual strands is of the order of tens of
microns, the use of such fine conductors in large systems as before leads to serious practical difficulties.

On the other hand, such "single-layer" conductors may form
a basis for the manufacture of completely transposed cables of
arbitrarily great cross section. The technology involved in making
such conductors has already been developed in the production of
various kinds of high-frequency multicore conductor (litz or strand
wire). First of all a plaited three-strand conductor has to be made
from the original conductors in the same way as a girl's hair is
plaited from three strands. (In principle the simple twisting of the
original conductors — possibly more than three — is sufficient;
however, such systems are clearly insufficiently reliable, since
the required symmetry may easily be infringed while twisting.)
From the resultant three-strand conductors thicker conductors
(nine-strand and so on) may then be plaited by the same means.

A slightly different technology may be applied in making plane
transposed conductors. First of all, using the original (single-
layer) conductors a tubular mesh is woven, and this is then crushed.
A large number of different arrangements may clearly be devised
for the production of transposed plaited conductors. The common
shortcoming of such systems is the low occupation factor associated with the large number of interweavings. In addition to this,
the fixing of the position of each individual filament in the overall
winding may be insufficiently rigid, so that displacements may
occur, and these are of course highly undesirable. These properties may be improved by mechanical compression, rolling,
impregnating with appropriate fillers, and so on.

On using plaited conductors in order to obtain constant magnetic fields it is immaterial whether the individual strands are
electrically connected or not (the question as to the time of penetration of the magnetic field into the interior of such a system is
as yet not taken into account). Hence the conductor used for this
purpose may be impregnated with a metal or alloy having appropriate properties. As well as ensuring the mechanical strength

of the cable, this metal or alloy may by virtue of its heat capacity, electrical conductivity, and other properties make a certain contribution toward the stabilization of the conductor with respect to flux jumps.

It is clear from the character of the field distribution in a completely transposed conductor that from the point of view of magnetization individual superconducting strands behave almost independently. The interaction of individual strands can in any case only lead to effects which we have entirely neglected in considering the set of strands as a continuous layer. Hence all the criteria of adiabatic stability proposed for single superconductors may in this case be directly used for determining the permissible parameters of the conductor. The current density in transposed conductors may therefore be far greater than in twisted conductors, despite the slightly smaller occupation factor. Special advantages of transposed conductors also appear in systems designed to produce rapidly varying magnetic fields (Section 11.4). All this makes transposed conductors a very promising material. The production of various types of such conductors, together with further study of and improvements to their properties, is accordingly extremely desirable.

Like the twisting of conductors, the transposition of individual strands has no effect at all on the mechanism of dynamic stabilization, and the criteria which we have obtained in the foregoing accordingly remain valid. Thus from the point of view of dynamic stability, transposition, like twisting, has no special significance.

11.4. Combined Conductors

for Obtaining Rapidly-Varying

Magnetic Fields

A detailed analysis of the behavior of combined superconducting materials in varying magnetic fields is not one of the tasks with which this book is concerned. Owing to the vast amount of existing theoretical and experimental material regarding this problem an exposition of the latter requires a special and much larger treatise. In this section we shall briefly consider some of the problems associated with calculating the losses in combined conductors.

The possibility of using superconductors to obtain strong, rapidly varying magnetic fields is being considered at the present time in connection with a large number of practical applications. The most grandiose prospect is that of constructing an annular synchrophasotron with a superconducting magnetic system. On increasing the induction in the working channel to 6-7 T (at the present time the value is approximately 1 T) there may be a proportionate reduction in the size of the whole ring for a specified maximum energy of the accelerated particles, or alternatively a still higher energy may be obtained. A large proportion of the difficulties associated with the realization of this effect is determined by the fact that in such an installation the field is required to increase from a certain (relatively small) initial value to its maximum in a time of the order of 1 sec. As we shall shortly see, this rate of field variation must be regarded as very high.

There are also other possibilities of using superconductors for obtaining "steady," rapidly varying fields in which effects associated with the varying nature of the field are not fundamental but second-order, and play a limiting, retarding role. As regards the possibility of obtaining essentially ac fields at, for example, industrial frequencies of 50-60 Hz, the level of the losses which may then be expected on using combined conductors of the kind at present available is extremely high. The efficiency of such systems, especially when compared with "ordinary" systems of obtaining alternating fields, is entirely inadequate.

Questions associated with the determination of losses in combined conductors also arise when calculating systems intended for producing a steady field but in which possible external actions lead to current fluctuations and field redistribution. An example of such a system may be mentioned in connection with plans to use superconducting magnetic systems for the noncontact suspension of transporting systems. In a number of such designs the lifting force has to be created by the interaction between magnets moving together with the transport and passive conductors situated in the roadbed. Any movements of the trucks relative to the roadbed arising from the unevenness of the latter, aerodynamic effects, and so on must inevitably lead ultimately to a change in both the current of the magnetic system and the magnetic field distribution.

The losses in a combined conductor associated with changes in the external field may be calculated if we know the character of the field distribution over the cross section of the conductor. Let us consider the model of a straight combined conductor, initially assuming the transport current to be equal to zero. Let the external field rise slowly enough for the field distribution at any given moment to be almost at equilibrium (see Section 9.2). Assuming the change in the field to be slight, we may regard the critical current density as being constant. For this case it is quite easy to obtain the following equation for the power of the mean specific losses in the conductor:

$$\dot{q} = \frac{1}{2b} \int_{-b}^{b} J_c K_s E \, dx = \frac{1}{2b} \int_{-b}^{b} K_s J_c \, d\xi \int_{0}^{\xi} \dot{B} x \, dx = \frac{1}{2} \mu_0 J_c K_s \dot{H} b. \quad (11\text{--}11)$$

Thus the density of the losses for a particular law of field penetration into the conductor is proportional to the thickness of the latter.

If the conductor carries a transport current I, the ratio of this current to the maximum critical current being ι so that $\iota = I/2bJ_c$, the integration in the expression for the losses should be carried out from the point of the "break" in the field distribution at which $E = 0$, and the final equation for the losses will assume the form

$$\dot{q} = \frac{1}{2}\mu_0 J_c K_s \dot{H}b(1 + \iota^2). \tag{11-12}$$

In order to determine the losses developing over the entire volume of the solenoid on increasing the field within the range in question, we may integrate the losses over the entire volume of the coil allowing for the specific J_c (B) relationship. Since ι is always less than unity for the whole range of field variation and in all parts of the coil except at the very center, we may use Eq. (11-11) for this calculation without introducing any serious error. It should be remembered that for weak fields (H < H_p) penetration is not complete. Owing to the rise in the current density in weak fields we must not neglect the part played by the regions corresponding to these fields. However, the detailed calculation is troublesome, since for any precise calculation of the losses the specific penetration law would be required.

We note that there is no very great point in making such an exact calculation for a straight combined conductor, since in systems intended to create pulsed fields completely transposed conductors would have to be used. Although the calculation of losses in the superconducting elements of transposed conductors is entirely in agreement with the calculations already presented, the presence of additional loss sources in these conductors also reduces the accuracy of such calculations.

Of greatest practical interest for the case of a straight conductor is the calculation of the losses taking place as a result of slight cyclical changes in an external field which penetrates into the conductor to only a shallow extent ($\Delta H < H_p$). The change in the field distribution for the case of zero transport current is shown in Fig. 11-6. Let us calculate the losses over that half of

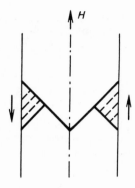

Fig. 11-6. Penetration of a cyclically varying external magnetic field into the conductor.

the cycle in which the external field diminishes by $2\Delta H$ from its maximum value:

$$\frac{1}{2}\,\Delta Q = 2 \int_0^{x_m} J_c \mu_0\,d\xi \int_0^\xi 2\Delta H\,\frac{x}{x_m}\,dx = \frac{2}{3}\,\mu_0\,\frac{(\Delta H)^3}{J_c}. \qquad (11\text{-}13)$$

The losses per cycle are thus

$$\Delta Q = \frac{4}{3}\,\mu_0 \left(\frac{\Delta H}{H_p}\right)^2 J_c b^2 \Delta H. \qquad (11\text{-}14)$$

This relationship may also be used for estimating the losses taking place in the initial magnetization of the conductor from zero to H_p (assuming J_c = const):

$$\Delta Q\,(H_p) = \frac{1}{3}\,\mu_0 J_c b^2 H_p. \qquad (11\text{-}15)$$

For sufficiently low transport currents, in which the changes in the field do not reach the "break" point in the initial field distribution, the foregoing equations clearly retain their validity. Since the line along which $E = 0$ still remains in its unaltered position in the center of the sample, no resistance to the transport current arises. The losses are thus covered entirely by the work of the source creating the variable field ΔH.

If, however, the amplitude of the change in the external field ΔH exceeds a certain threshold value, determined by the condition for the varying field to penetrate into the conductor by an amount

$b(1 - I/2bJ_c)$ so as to reach the "break" point, the picture changes considerably [103]. We shall not present any analysis here of the losses in the material taking place after the amplitude ΔH has exceeded the threshold value ΔH_{th}, where

$$\Delta H_{th} = J_c b \left(1 - \frac{I}{2bJ_c} \right). \tag{11-16}$$

We simply note that some of the losses will now be covered by the work of the transport-current source. This may be of no small importance in systems required to operate under short-circuit conditions. If as a result of external actions the amplitude of the field increment at any point of the coil exceeds ΔH_{th}, the current in the system will start diminishing.

Let us consider the question of calculating the losses in twisted and transposed conductors. In the absence of a transport current the external field will penetrate completely into the normal regions of the conductor, and for sufficiently slow changes of field the losses in these regions may be reduced to negligibly small values. However, the losses associated with the finite magnetization of the superconductor remain intact. In order to determine these, we easily see that Eq. (11-11) may be used, after replacing b on its right-hand side by the radius of the superconducting strands r_s (or half the thickness of the superconducting layer in a strip):

$$q = \frac{1}{2} \mu_0 K_s r_s \overline{J_c \Delta H}; \tag{11-17}$$

here $\overline{J_c \Delta H}$ represents a certain average value taken over the cross section of the conductor.

It is quite clear that for completely transposed conductors carrying a transport current Eq. (11-11) may once again be employed, since it is permissible to regard the individual strands of the superconductor as independent, and since the proportion ι of the transport current in each of these is equal to the same quantity for the conductor as a whole. Thus,

$$q = \frac{1}{2} \mu_0 K_s r_s \overline{J_c \Delta H} (1 + \iota^2). \tag{11-18}$$

In a twisted conductor with a constant transport current $I = \iota 2bJ_c$ a proportion ι of the total number of strands carry the critical current, and the losses for these are determined by Eq. (11-18) with the value of ι in this equation set equal to unity. For the remaining $(1 - \iota)$ we may use Eq. (11-17), since the transport current does not penetrate into these strands. Finally we have

$$q = \frac{1}{2} \mu_0 K_s r_s \overline{J_c \, \Delta H} (1 + \iota). \tag{11-19}$$

Here we are not considering the more complicated cases in which the transport current changes together with the external field or independently of the latter. In view of the smallness of ι for the whole range of fields and coil volumes, the contribution of such combined effects to the total losses during the complete magnetization of the system is quite small.

We simply note that, when the transport current in twisted conductors changes, owing to the fact that the field of the current does not penetrate completely into the conductor, additional losses arise, the maximum average density Q_c of these being given by the equation [93]

$$Q_c \approx \mu_0 K_s J_c b H_{int}, \tag{11-20}$$

where H_{int} is the maximum "intrinsic" field of the conductor: $H_{int} = bJ_c K_s/2$. It is clear that these additional losses are only determined by the total size of the conductor and therefore cannot be reduced by making individual strands smaller. Thus in principle we may have to impose a certain limitation on the maximum diameter of the conductor, for which the losses Q_c become comparable with the losses in the superconductor itself (for specified strand dimensions). On the other hand, for a fixed external size we may in this way deduce the minimum dimensions of the strands, such that any further "fractioning" of the superconductor becomes pointless. However, the exact calculation of such a limiting condition is extremely difficult [93].

It should also be remembered that in calculating the losses during the magnetization of extremely fine superconducting strands (a few microns in diameter) the foregoing type of calculation may

be too approximate. In this limiting case we should be required
to make a more precise allowance for the thermodynamic and magnetic properties of the superconductor, and also various surface
effects; hence the numerical values of the losses determined by
means of equations such as (11-11) should be treated as extremely
approximate.

Up to the present time no allowance has been made for losses
in the normal regions of the combined conductor, since it has been
assumed that the change in field takes place quite slowly. However,
we are here interested in the greatest possible rate of field increment, and we must therefore now estimate the losses arising
in the case of finite rates of change. We shall of course only consider processes slow enough for their characteristic times to
exceed the time τ_0 required for the longitudinal penetration of the
external field into the twisted or transposed conductor. For high
rates of field increment the field penetrates into the conductor in
the transverse direction, i.e., independently of whether the conductor is twisted or not. The loss level then corresponds to the losses
in a straight conductor, which, as may easily be shown on the basis
of the foregoing relationships, may be quite unacceptable.

Let us first consider the losses taking place for a fairly slow
penetration of the external field into the twisted conductor. Let us
once again turn to the "inverted" model with a straight conductor
and a "twisted" field (Fig. 11-1a). It is clear that we may again
consider the penetration of only one component of the external
field into the conductor; in order to obtain the final expression we
may then simply double the resultant loss power. The penetration
of one field component will take place in approximately the same
way as into a plane layer of normal metal enclosed between two
superconducting planes (Fig. 11-1b). The width 2b of such a layer
should be approximately equal to the mean distance between the
individual superconducting strands in the conductor.

The losses in such a layer of normal metal when the external
magnetic field changes may be calculated quite accurately. However, remembering the many simplifications which we have had
to make in constructing the model, we may again confine attention
to a simple rough estimate. For this purpose we may replace the
layer of metal considered by an equivalent circuit containing an
inductance and an active impedance (resistance). The equivalent

inductance may be written in the form

$$L = \frac{\Psi}{I} \approx \frac{SB}{I} \approx \mu_0 \frac{l_0 b}{2h}, \tag{11-21}$$

where Ψ is the magnetic flux threading the circuit; S is the equivalent area of the circuit around which the current is flowing, approximately equal to $^1/_4 \, l_0 \times 2b$; l_0 is the twist pitch; and h is the height of the layer.

The resistance R of the circuit under consideration arises when the current I flows through the normal metal between the layers of superconductor

$$R \approx 16 \, \frac{\rho_n b}{l_0 h}. \tag{11-22}$$

It follows from this that the time constant of the circuit equals

$$\tau_0 = L/R \approx \frac{\mu_0 \, l_0^2}{32 \rho_n}, \tag{11-23}$$

which practically agrees with Eq. (9-6), according to which the factor $^1/_{32}$ is replaced by $^1/_4 \pi^2 \approx \, ^1/_{40}$ (considering that b = $l_0/4$).

For the fairly slow processes here considered, in which the characteristic time is much greater than τ_0, the external field is able to penetrate completely into the layer. Hence in order to determine the losses we may use the following relationship:

$$\dot{Q} = \frac{U^2}{R} \approx \frac{(SB)^2}{R}, \tag{11-24}$$

where U is the voltage in the equivalent circuit.

Using this equation, let us determine the total losses in the normal metal during the magnetization of the conductor from zero field to a certain value H_m. Let us assume that the field increases in accordance with the law

$$H = H_m (1 - e^{-t/\tau}). \tag{11-25}$$

Integrating the power of the losses in (11-24) with respect to time we obtain

$$Q = \int\limits_{0}^{\infty} \dot{Q}\,dt = \frac{S^2 B_m^2}{R\tau^2} \int\limits_{0}^{\infty} 2t/\tau \; dt = \frac{\tau_0}{\tau} \frac{\mu_0 \operatorname{sh} H_m^2}{2} \approx W_m \frac{\tau_0}{\tau}. \qquad (11\text{-}26)$$

Thus the ratio of the total losses to the total energy of the magnetic field roughly equals τ_0/τ.

We note that the energy of the field penetrating into the superconducting layers was not taken into account, since for purposes of simplification it was assumed that the depth of penetration was much smaller than the thickness of the normal layers. This assumption is only completely acceptable for small occupation factors K_s. In general we should also ascribe a certain proportion of the thickness of the superconductor to the thickness of the normal region, since the magnetic flux, rather than penetrate into the superconductor, should intersect the complete layer of normal metal in a longitudinal direction. Hence in estimating the losses we should allow for the energy of the magnetic field over the entire volume of the conductor and not simply the normal regions. In addition to this, the intrinsic penetration time τ_0 should also in general be increased by about $(1 + K_s/K_{tot})$ times, since the equivalent inductance allowing for the flux in the superconductor increases in approximately this ratio.*

The accuracy of the estimates should be slightly higher for the case of small oscillations of the field around a specific value, especially if the depth of penetration of the varying component into the superconductor is slight by comparison with the diameter of the strand. If the field varies in accordance with the law $H = H_0 + H_1 \sin \omega t$, then by using Eq. (11-24) we may obtain the following equation for the power of the losses in the normal metal:

$$\dot{Q} = \frac{\overline{U^2}}{R} = \int\limits_{0}^{\frac{2\pi}{\omega}} \frac{U^2}{R}\,dt \approx W_1 \tau_0 \omega^2, \qquad (11\text{-}27)$$

where W_1 is the "magnetic energy" of the varying component of the field, equal to $\mu_0 H_1^2 Sh/2$.

As in the previous case, for large amplitudes H_1 the quantities τ_0 and W_1 should really be referred to the total volume of the con-

---

*These corrections are of course not introduced very accurately here, since it is not entirely consistent even to separate the losses in the normal metal from those in the superconductor.

ductor, and the accuracy of the estimates will again be reduced in
this case.

By using Eq. (11-26) or (11-27) we may approximately estimate the permissible rate of field growth or frequency (and amplitude) of the variable field component, considering, for example,
that the additional losses in the normal metal become equal to the
losses in the superconductor at this frequency (growth rate). It
is clear that the maximum possible growth rate (frequency of variation) of the field is determined by the intrinsic time τ_0 of the
conductor, since in faster processes the field will penetrate not
along the conductor but over its radius, i.e., in the same way as
in an untwisted conductor.

In order to characterize the permissible rate of variation of
the external field, the concept of the critical twist pitch is often
introduced. Corresponding to this critical pitch there is a certain specified rate of change of the external field $\Delta H / \Delta t$, for which
the difference in the field values on opposite sides of the superconducting strand in the outer layer of the conductor is equal to
$2bJ_c$, i.e., when the current in this strand corresponds to the critical
value. Of course for greater rates of field change the superconductivity of the outer strands will be destroyed by the current, and
penetration of the field will take place across the direction of the
conductor. Thus for the critical pitch the permissible rate of
field variation is approximately equal to dJ_c / τ_0, where d is the
half width of the conductor, while τ_0 is determined from the second part of Eq. (9-6).

In transposed conductors the loss characteristics in the normal
metal mainly correspond to their characteristics in twisted conductors, although the loss level is largely determined by the set of
penetration times, the greatest of which corresponds to the complete transposition pitch. The complete transposition pitch is the
name given to the distance along the axis of the conductor within
which an individual strand, having passed through all the positions
occupied by the other strands, returns to its original state. Owing
to the extremely complicated geometrical structure of the transposed conductor, any detailed calculation of the losses is very difficult.

For any specific combined conductor the losses in a varying
field may be quite reliably estimated by measuring the magnetization of the conductor in the range of fields and amplitudes and

frequencies of the varying component presenting the greatest current interest. The losses are proportional to the area of the hysteresis loop in the M(H) relationship. The results of such measurements [93] on the whole quite accurately confirm the simple considerations and loss estimates just presented.

In order to reduce the intrinsic time of field penetration into combined conductors intended to produce varying fields, we may use an alloy with a low electrical conductivity as matrix material. However, the use of such conductors in any large magnetic systems involves serious difficulties, since in the case of a chance transition into the normal state the coil of the system will be damaged (see Section 3.1). We find in addition that such conductors are unusually sensitive to possible displacements of the turns inside the winding, since without reliable impregnation the critical currents in the windings are extremely low. This is evidently associated with the fact that the outflow of heat from the deformed parts of the combined conductor is impeded when there is no efficient heat-conducting coating.

More convenient than this are three-component materials incorporating high-purity copper and alloys with a low electrical conductivity in addition to the superconducting strands. A description of some combined conductors of this kind was presented in [93, 99], each superconducting strand being encased in a copper—nickel alloy sheath. The general matrix of the conductor was made of copper. This structure was no doubt adopted for technological reasons. A rather different disposition of the conductors is better from the point of view of both reducing the losses and making a more rational use of the heat capacity of the components; in this each strand is encased in a copper sheath and the general matrix is made of the alloy. We note that the safety of both types of conductor is simply determined by the total amount of copper and does not depend on the manner in which this is distributed over the cross section.

A completely transposed (safe in operation) conductor with a reduced intrinsic field penetration time τ_0 can be made by using thin twisted isolated wires with a copper matrix as original material for the plaiting. Every such wire may contain more than three superconducting strands arranged in a single layer. Chance short circuitings of individual wires which may occur as a result of the imperfection of the insulation (provided that these are not too great in number) should have hardly any effect on the total losses in the conductor. For calculating the losses in the normal regions of such conductors we may make use of Eq. (11-27), which allows for the magnetic energy relating to an individual wire and the intrinsic time determined by the twist pitch of the latter.

As already mentioned, plaited transposed conductors are also extremely sensitive to possible displacements of individual strands, and should therefore be provided with appropriate impregnation. This requirement clearly contradicts the requirement that heat evolved as a result of the changes in field should be carried away from the coil. In a correctly made coil uniformly distributed helium channels should accordingly be provided. Such a coil should also clearly contain additional elements to take up the mechanical stresses which arise. A detailed discussion of these problems has been presented by Smith and Lewin [104].

APPENDIX

Dimensionless Volt–Ampere Characteristics
of Combined Conductors

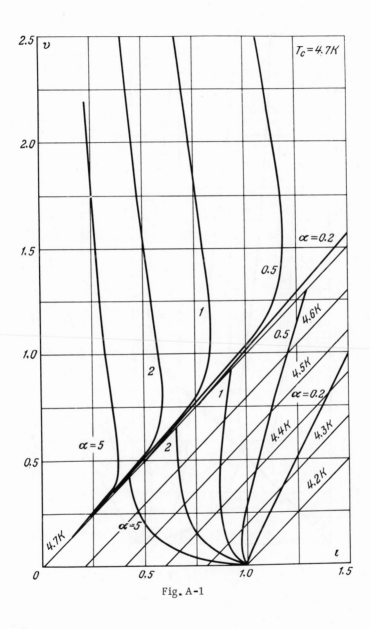

Fig. A-1

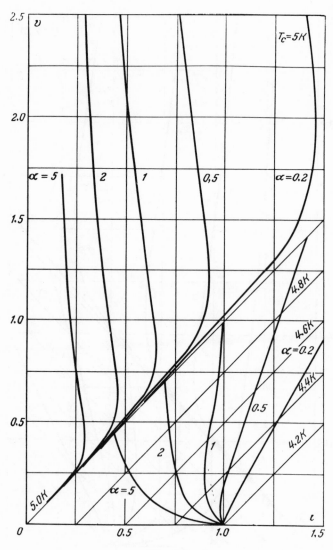

Fig. A-2

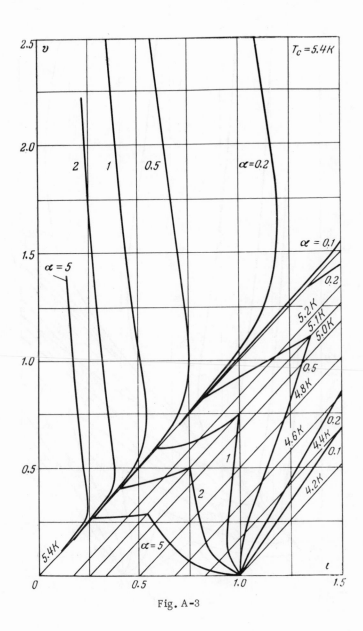

Fig. A-3

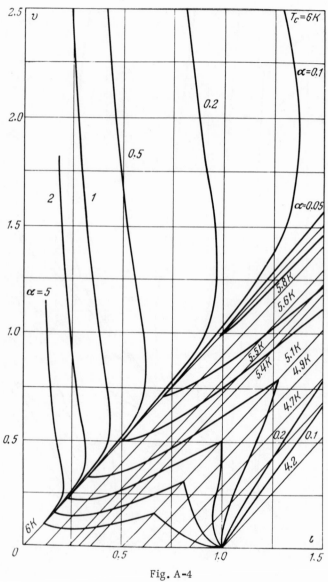

Fig. A-4

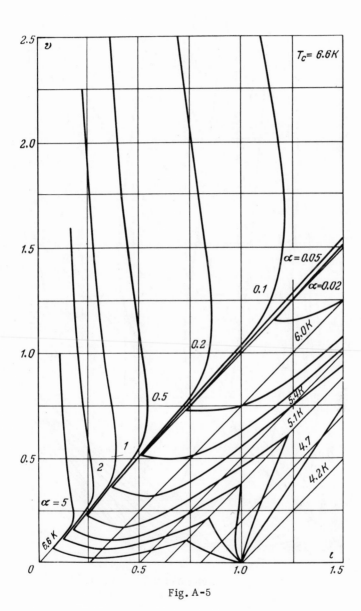

Fig. A-5

332

APPENDIX

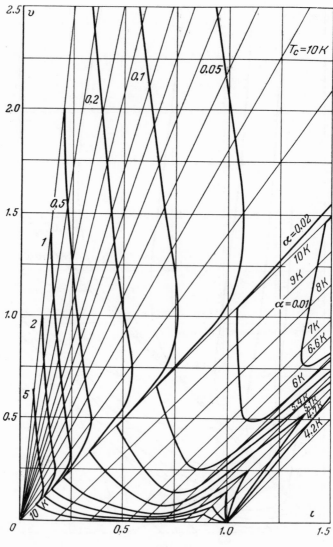

Fig. A-6

References

1. H. Kamerlingh-Onnes, Commun. Phys. Lab. Univ. Leiden, No. 120b (1911).
2. H. Kamerlingh-Onnes, Commun. Phys. Lab. Univ. Leiden, No. 122b (1911).
3. H. Kamerlingh-Onnes, Commun. Phys. Lab. Univ. Leiden, No. 133d (1913) and No. 134f (1914).
4. D. B. Montgomery, Strong Magnetic Fields — Production and Application, CNRS, Paris (1967).
5. V. G. Veselago, L. P. Maksimov, and A. M. Prokhorov, Pribory i Tekh. Eksperim., 1968(4):192 [English translation: Instrum. Exp. Tech. (USSR), 1968:967].
6. W. Samspon, P. P. Craig, and M. Strongin, Sci. Amer., 216:115 (1967).
7. R. H. Kropschot and V. Arp, Cryogenics, 2:1 (1961).
8. P. F. Chester, Report on the Royal Society, Meeting for Discussion of Advanced Methods of Energy Conversion — Magnetohydrodynamic Power Generation, London (November 4, 1965).
9. D. E. Spiel and R. W. Boom, Low Temperatures and Electric Power, Int. Conf., London (March 24-26, 1969), p. 161.
10. M. Ferrier, ibid., p. 150.
11. V. B. Zenkevich and V. V. Sychev, Superconducting Magnetic Systems [in Russian], Nauka, Moscow (1972).
12. V. V. Sychev and V. B. Zenkevich, Atomic Energy Review, Vol. 11, No. 4 (1974).
13. H. London and F. London, Proc. Roy. Soc., A149:71 (1935); Physica, 2:341 (1935).
14. V. V. Sychev, Complex Thermodynamic Systems [in Russian], Énergiya, Moscow (1970).
15. J. Bardeen, L. N. Cooper, and J. Schrieffer, Phys. Rev., 108:1175 (1957).
16. L. N. Cooper, Phys. Rev., 104:1189 (1956).
17. V. L. Ginzburg and L. D. Landau, Zh. Éksp.Teor. Fiz., 20:1064 (1950).
18. A. A. Abrikosov, Zh. Éksp. Teor. Fiz., 32:1442 (1957).
19. L. P. Gor'kov, Zh. Éksp.Teor. Fiz., 34:735 (1958); 36:1918 (1959); 37:833, 1407 (1959).
20. F. London, Superfluids, Vol. 1, John Wiley and Sons, New York (1950).
21. R. D. Parks, "Quantum effects in superconductors," in Quantum Macrophysics [in Russian], Nauka, Moscow (1967).

22. V. L. Ginzburg, Superconductivity [in Russian], Izd. AN SSSR, Moscow (1946).
23. A. M. Glogston, Phys. Rev. Letters, 2:266 (1962).
24. P. G. De Gennes, Superconductivity of Metals and Alloys [English translation], W. A. Benjamin, New York(1966).
25. C. Laverick, Superconducting Magnets [Russian translation], Mir, Moscow (1968).
26. J. Silcox and R. W. Rollins, Rev. Mod. Phys., 36:52 (1964).
27. Y. B. Kim, C. F. Hempstead, and A.R. Strnad, Phys. Rev., 129:528 (1963).
28. Y. B. Kim, C. F. Hempstead, and A.R. Strnad, Phys. Rev., 131:2486 (1963).
29. Y. B. Kim, Phys. Today (September 21, 1964).
30. P. W. Anderson and Y. B. Kim, Rev. Mod. Phys., 36:62 (1964).
31. C. J. Gorter, Physica, 31:407 (1965).
32. A. R. Strnad, C. F. Hempstead, and Y.B. Kim, Phys. Rev. Letters, 13:794 (1964).
33. Y. B. Kim, C. F. Hempstead, and A.R. Strnad, Phys. Rev., 139:A1163 (1965).
34. H. C. Schindler, J. Appl. Phys., 40:2110 (1969).
35. V. V. Andrianov, V. B. Zenkevich, V. I. Sokolov, V. V. Sychev, V. A. Tovma, and L. N. Fedotov, Dokl. Akad. Nauk SSSR, 169:316 (1966) [English translation: Sov. Phys. Dokl., 11:619 (1966)].
36. V. A. Kirillin, V. V. Sychev, and A. E. Sheindlin, Technical Thermodynamics [in Russian], Énergiya, Moscow (1968).
37. V. V. Sychev, V. B. Zenkevich, and V. V. Andrianov, Izv. Akad. Nauk SSSR, Énergetika i Transport, No. 1, p. 100 (1965).
38. V. V. Sychev, V. B. Zenkevich, and V. V. Andrianov, Izv. Akad. Nauk SSSR, Énergetika i Transport, No. 2, p. 117 (1965).
39. V. V. Sychev, V. B. Zenkevich, and V. V. Andrianov, Electricity from MHD, Proc. Symposium, Salzburg, Vol. III, IAEA, Vienna (1966).
40. M. W. Dowley, Cryogenics, 4:153 (1964).
41. D. L. Watrons, IEEE Trans. on Magnetics, MAG-1:402 (1965).
42. T. G. Berlincourt, Brit. J. Appl. Phys., 14:749 (1963).
43. P. F. Smith et al., Proc. 1968 Summer Study of Superconducting Devices, BNL (1969).
44. R. Hancox, IEEE Trans. on Magnetics, MAG-4:486 (1968).
45. P. F. Chester, 1967 Int. Cryog. Eng. Conf. Japan, Rep. 111-12a1 (April 9-13, 1967).
46. E. R. Schrader, J. Appl. Phys., 40:2016 (1969).
47. Y. Iwasa, C. Weggel, D. B. Montgomery, R. Weggel, and F. R. Hales, J. Appl. Phys., 40:2006 (1969).
48. A. R. Kantorovitz and Z. J. J. Stekly, Appl. Phys. Letters, 6:56 (1965).
49. F. Pawlek and D. Rogalla, Cryogenics, 6:14 (1966).
50. Westinghouse Technical Data (1963), pp. 53-161; (1964), p. 29/64A-64637.
51. L.Donadieu andJ.Maldy, Bull.d'Inform.Scient.-Techn. du Commissariat à l'Energie Atomique, No. 108 (1968).
52. V. V. Sychev and V. A. Al'tov, Izv. Akad. Nauk SSSR, Énergetika i Transport, No. 5 (1970).
53. Z. J. J. Stekly, Proc. Summer Study Superconducting Devices, BNL (1969).
54. M. G. Kremlev, Cryogenics, 7:267 (1967).

55.   H. Brechna, Proc. Summer Study Superconducting Devices, BNL (1969).
56.   B. J. Maddock, G. B. James, and W. T. Norris, Cryogenics, 9:261 (1969).
57.   M. G. Kremlev, V. B. Zenkevich, and V. A. Al'tov, Cryogenics, 8:173 (1968).
58.   R. D. Cummings and J. L. Smith, Liquid Helium Technology, Pergamon Press, London (1966), p. 85.
59.   V. E. Keilin et al., Strong Magnetic Fields — Production and Application, CNRS, Paris (1967).
60.   T. M. Dauphinee and H. Preston-Thomas, Rev. Sci. Instr., 25:884 (1954).
61.   V. V. Sychev, V. B. Zenkevich, M. G. Kremlev, and V. A. Al'tov, Dokl. Akad. Nauk SSSR, 188:83 (1969) [English translation: Sov. Phys. Dokl., 14:911 (1969)].
62.   V. V. Sychev, V. B. Zenkevich, and V. A. Al'tov, Zh. Tekh. Fiz., 42:1288 (1972) [English translation: Sov. Phys. Tech. Phys., 17:1023 (1972)].
63.   W. F. Gauster and J. B. Hendricks, IEEE Trans. on Magnetics, MAG-4:489 (1968).
64.   C. Whetstone and R. W. Boom, Adv. Cryog. Eng., 13:68 (1968).
65.   M. S. Lubell and D. M. Kroger, Adv. Cryog. Eng., 14:123 (1969).
66.   V. V. Sychev et al., Izv. Akad. Nauk SSSR, Énergetika i Transport, No. 4, p. 88 (1972).
67.   Z. J. J. Stekly, J. Appl. Phys., 37:324 (1966).
68.   W. F. Gauster and J. B. Hendricks, J. Appl. Phys., 39:2572 (1968).
69.   V. V. Sychev, V. B. Zankevich, V. A. Al'tov, M. G. Kremlev, and N. A. Kulysov, Cryogenics, 12:377 (1972).
70.   V. V. Sychev and V. A. Al'tov, Dokl. Akad. Nauk SSSR, 199:817 (1971) [English translation: Sov. Phys. Dokl., 16:673 (1971)].
71.   M. N. Wilson, Liquid Helium Technology, Pergamon Press, London (1966), p. 109.
72.   W. F. Gauster, J. Appl. Phys., 40:2060 (1969).
73.   S. Ya. Berkovich and M. G. Kremlev, Transactions of the All-Union Conference on Computing Systems [in Russian], Inst. Math., Novosibirsk (1968), p. 88.
74.   V. V. Sychev, B. V. Zankevich, V. A. Al'tov, M. G. Kremlev, and N. A. Kulysov, Cryogenics, 13:19 (1973).
75.   A. P. Smirnov, I. S. Parshina, T. A. Rusanova, and V. N. Totubalin, Zh. Tekh. Fiz., 38:1588 (1968) [English translation: Sov. Phys. Tekh. Phys., 13:1290 (1968)].
76.   V. V. Sychev, V. B. Zenkevich, and V. A. Al'tov, Izv. Akad. Nauk SSSR, Énergetika i Transport, No. 6 (1970).
77.   F. Pawlek and D. Rogalla, Cryogenics, 6:12 (1968).
78.   V. A. Al'tov, M. G. Kremlev, V. V. Sychev, and V. B. Zenkevich, Cryogenics, 13:420 (1973).
79.   S. Shimamoto and H. Desportes, J. Appl. Phys., 41:3286 (1970).
80.   M. P. Malkov et al., Handbook on the Physico-technical Foundations of Cryogenics [in Russian], Énergiya, Moscow (1973).
81.   M. Morpurgo, "Construction of a superconducting test coil cooled by helium forced circulation," CERN 68-17, Geneva (1968).
82.   V. E. Keilin, E. Yu. Klimenko, and I. A. Kovalev, Cryogenics, 9:36 (1969).
83.   V. E. Kelin, and I. A. Kovalev, Cryogenics, 9:100 (1969).
84.   W. J. Greene, Adv. Cryog. Eng., 13:138 (1969).

85. V. E. Keilin et al., Cryogenics, 12:292 (1972).
86. Z. J. J. Stekly, J. Appl. Phys., 37:324 (1966).
87. R. W. Boom and L. D. Roberts, J. Appl. Phys., 34:2422 (1963).
88. Z. J. J. Stekly, Adv. Cryog. Eng., 8:585 (1963).
89. V. V. Sychev, V. B. Zenkevich, and V. A. Al'tov, Izv. Akad. Nauk SSSR, Énergetika i Transport, No. 6, p. 100 (1970).
90. P. F. Chester, Brit. Patent No. 1124622 (1964).
91. P. F. Smith, Proc. Second Int. Conf. Magnet. Technol., Oxford (1967).
92. F. Lange, Cryogenics, 6:176 (1966).
93. M. N. Wilson, C. R. Walters, J. D. Lewin, P. F. Smith, and A. H. Spurway, J. Phys. D, 3:1517 (1970).
94. S. L. Wipf, Phys. Rev., 161:404 (1967).
95. P. S. Swartz and C. P. Bean, J. Appl. Phys., 39:4991 (1968).
96. J. D. Livingstone, Appl. Phys. Letters, 8:319 (1966).
97. Y. Iwasa and D. B. Montgomery, Adv. Cryog. Eng., 14:114 (1969).
98. R. Hancox, IEEE Trans. on Magnetics, MAG-4:486.
99. A. C. Barber and P. F. Smith, Cryogenics, 9:483 (1969).
100. P. F. Smith, A. H. Spurway, and J. D. Lewin, Brit. J. Appl. Phys., 16:947 (1965).
101. P. F. Chester, Rep. Prog. Phys., 30:561 (1967).
102. H. R. Hart, Proc. 1968 Summer Study on Superconducting Devices, BNL (1969), p. 571.
103. V. V. Andrianov, V. B. Zenkevich, V. V. Kurguzov, V. V. Sychev, and F. F. Ternovskii, Zh. Éksp. Teor. Fiz., 58:1523 (1971) [English translation: Sov. Phys. JETP, 31:815 (1971)].
104. P. F. Smith and J. D. Lewin, Nucl. Instrum. Methods, 52:298 (1967).

Index

131895